Practical Troubleshooting of Electrical Equipment and Control Circuits

Other titles in the series

Practical Cleanrooms: Technologies and Facilities (David Conway)

Practical Data Acquisition for Instrumentation and Control Systems (John Park, Steve Mackay)

Practical Data Communications for Instrumentation and Control (Steve Mackay, Edwin Wright, John Park)

Practical Digital Signal Processing for Engineers and Technicians (Edmund Lai)

Practical Electrical Network Automation and Communication Systems (Cobus Strauss)

Practical Embedded Controllers (John Park)

Practical Fiber Optics (David Bailey, Edwin Wright)

Practical Industrial Data Networks: Design, Installation and Troubleshooting (Steve Mackay, Edwin Wright, John Park, Deon Reynders)

Practical Industrial Safety, Risk Assessment and Shutdown Systems for Instrumentation and Control (Dave Macdonald)

Practical Modern SCADA Protocols: DNP3, 60870.5 and Related Systems (Gordon Clarke, Deon Reynders)

Practical Radio Engineering and Telemetry for Industry (David Bailey)

Practical SCADA for Industry (David Bailey, Edwin Wright)

Practical TCP/IP and Ethernet Networking (Deon Reynders, Edwin Wright)

Practical Variable Speed Drives and Power Electronics (Malcolm Barnes)

Practical Centrifugal Pumps (Paresh Girdhar and Octo Moniz)

Practical Electrical Equipment and Installations in Hazardous Areas (Geoffrey Bottrill and G. Vijayaraghavan)

Practical E-Manufacturing and Supply Chain Management (Gerhard Greef and Ranjan Ghoshal)

Practical Grounding, Bonding, Shielding and Surge Protection (G. Vijayaraghavan, Mark Brown and Malcolm Barnes)

Practical Hazops, Trips and Alarms (David Macdonald)

Practical Industrial Data Communications: Best Practice Techniques (Deon Reynders, Steve Mackay and Edwin Wright)

Practical Machinery Safety (David Macdonald)

Practical Machinery Vibration Analysis and Predictive Maintenance (Cornelius Scheffer and Paresh Girdhar)

Practical Power Distribution for Industry (Jan de Kock and Cobus Strauss)

Practical Process Control for Engineers and Technicians (Wolfgang Altmann)

Practical Power Systems Protection (Les Hewitson, Mark Brown and Ben. Ramesh)

Practical Telecommunications and Wireless Communications (Edwin Wright and Deon Reynders)

Practical Hydraulics (Ravi Doddannavar, Andries Barnard)

Practical Batch Process Management (Mike Barker, Jawahar Rawtani)

Practical Troubleshooting of Electrical Equipment and Control Circuits

Mark Brown Pr.Eng, DipEE, B.Sc (Elec Eng),
Senior Staff Engineer, IDC Technologies, Perth, Australia

Jawahar Rawtani M.Sc (Tech), MBA,
Senior Electrical Engineer, Nashik, India

Dinesh Patil BEng,
Patil and Associates.

Series editor: Steve Mackay FIE (Aust), CPEng, B.Sc (Elec.Eng), B.Sc (Hons), MBA,
Gov.Cert.Comp., Technical Director – IDC Technologies

AMSTERDAM • BOSTON • HEIDELBERG • LONDON
NEW YORK • OXFORD • PARIS • SAN DIEGO
SAN FRANCISCO • SINGAPORE • SYDNEY • TOKYO

Newnes is an imprint of Elsevier

Newnes

Newnes is an imprint of Elsevier
The Boulevard, Langford Lane, Kidlington, Oxford, OX5 1GB
30 Corporate Drive, Suite 400, Burlington, MA 01803, USA

First edition 2005
Reprinted 2007, 2008, 2009

Notice
No responsibility is assumed by the publisher for any injury and/or damage to persons
or property as a matter of products liability, negligence or otherwise, or from any use
or operation of any methods, products, instructions or ideas contained in the material
herein. Because of rapid advances in the medical sciences, in particular, independent
verification of diagnoses and drug dosages should be made

British Library Cataloguing in Publication Data
A catalogue record for this book is available from the British Library

Library of Congress Cataloging-in-Publication Data
A catalog record for this book is available from the Library of Congress

ISBN: 978-0-7506-6278-9

For information on all Newnes publications
visit our website at www.elsevierdirect.com

Printed and bound in the United Kingdom

Transferred to Digital Print 2010

Working together to grow
libraries in developing countries

www.elsevier.com | www.bookaid.org | www.sabre.org

ELSEVIER   BOOK AID
International   Sabre Foundation

Contents

Preface

There is a large gap between the theory of electron flow, magnetic fields and that of troubleshooting electrical equipment and control circuits in the plant. In this book, we try to avoid or at least minimize discussions on the theory and instead focus on showing you how to troubleshoot electrical equipment and control circuits. The book helps to increase your knowledge and skills in improving equipment productivity whilst reducing maintenance costs. Reading this book will help you identify, prevent and fix common electrical equipment and control circuits. The focus is 'outside the box'. The emphasis is on practical issues that go beyond typical electrical theory and focus on providing those that attend with the necessary toolbox of skills in solving electrical problems, ranging from control circuits to motors and variable speed drives.

This book focuses on the main issues of troubleshooting electrical equipment and control circuits of today to enable you to walk onto your plant or facility to troubleshoot and fix problems as quickly as possible. This is not an advanced book but one aimed at the fundamentals of troubleshooting systems. The book is very practical in its approach to troubleshooting and the examples you will be shown are applicable to any facility.

We would hope that you will gain the following knowledge from this book:

- Diagnose electrical problems 'right-first-time'
- Minimize the expensive trial and error troubleshooting approach
- Reduce unexpected downtime on electrical motors and other equipment
- Improve plant safety
- Learn specific techniques to troubleshoot equipment and control circuits
- Analyze equipment problems
- Determine causes of equipment failure
- Troubleshoot electrical equipment and control circuits.

Typical people who will find this book useful include:

- Plant Electricians
- Mechanical Engineers
- Production Operators and Supervisors
- Utilities Maintenance Personnel
- Plant Engineers.

Pre-requisites

A basic understanding of electrical theory and problems you have encountered in the past would be helpful but a basic review is undertaken at the beginning of the book.

1

Basic principles

Objectives

- To refresh basic electrical concepts
- To define basic concepts of transformer
- To refresh single-phase power concepts
- To refresh three-phase power concepts.

1.1   Introduction

A significant proportion of industrial electricity is about single-phase and three-phase transformers, AC and DC machines. In this context, we will study the electrical circuits and their construction, design, testing, operation, and maintenance.

For troubleshooting electrical equipment and control circuits, it is important to understand the basic principles on which the electrical equipment works. The following sections will outline the basic electrical concepts.

1.1.1   Basic electrical concepts

In each plant, the mechanical movement of different equipments is caused by an electric prime mover (motor). Electrical power is derived from either utilities or internal generators and is distributed through transformers to deliver usable voltage levels.

Electricity is found in two common forms:

- AC (alternating current)
- DC (direct current).

Electrical equipments can run on either of the AC/DC forms of electrical energies. The selection of energy source for equipment depends on its application requirements. Each energy source has its own merits and demerits.

Industrial AC voltage levels are roughly defined as LV (low voltage) and HV (high voltage) with frequency of 50–60 Hz.

An electrical circuit has the following three basic components irrespective of its electrical energy form:

- Voltage (volts)
- Ampere (amps)
- Resistance (ohms).

Voltage is defined as the electrical potential difference that causes electrons to flow. Current is defined as the flow of electrons and is measured in amperes.

Resistance is defined as the opposition to the flow of electrons and is measured in ohms.

All three are bound together with Ohm's law, which gives the following relation between the three:

$$V = I \times R$$

Where
  V = Voltage
  I = Current
  R = Resistance.

(a) *Power*

In DC circuits, power (watts) is simply a product of voltage and current.

$$P = V \times I$$

For AC circuits, the formula holds true for purely resistive circuits; however, for the following types of AC circuits, power is not just a product of voltage and current.

Apparent power is the product of voltage and ampere, i.e., VA or kVA is known as apparent power. Apparent power is total power supplied to a circuit inclusive of the true and reactive power.

Real power or true power is the power that can be converted into work and is measured in watts.

Reactive power   If the circuit is of an inductive or capacitive type, then the reactive component consumes power and cannot be converted into work. This is known as reactive power and is denoted by the unit VAR.

(b) *Relationship between powers*

$$\text{Apparent power (VA)} = V \times A$$

$$\text{True power (Watts)} = \text{VA} \times \cos \phi$$

$$\text{Reactive power (VAR)} = \text{VA} \times \sin \phi$$

(c) *Power factor*

Power factor is defined as the ratio of real power to apparent power. The maximum value it can carry is either 1 or 100(%), which would be obtained in a purely resistive circuit.

$$\text{Power factor} = \frac{\text{True power}}{\text{Apparent power}}$$

$$\frac{\text{Watts}}{\text{kVA}}$$

(d) *Percentage voltage regulation*

$$\% \text{ Regulation} = 100 \frac{(\text{No load voltage} - \text{Full load voltage})}{\text{Full load voltage}}$$

(e) *Electrical energy*

This is calculated as the amount of electrical energy used in an hour and is expressed as follows:

$$\text{Kilowatthour} = kW \times h$$

Where
  kW = kilowatt
  h = hour.

(f) *Types of circuits*

There are only two types of electrical circuits – series and parallel.

A series circuit is defined as a circuit in which the elements in a series carry the same current, while voltage drop across each may be different.

A parallel circuit is defined as a circuit in which the elements in parallel have the same voltage, but the currents may be different.

1.1.2   Transformer

A transformer is a device that transforms voltage from one level to another. They are widely used in power systems. With the help of transformers, it is possible to transmit power at an economical transmission voltage and to utilize power at an economic effective voltage.

Basic principle

Transformer working is based on mutual *emf* induction between two coils, which are magnetically coupled.

When an AC voltage is applied to one of the windings (called as the primary), it produces alternating magnetic flux in the core made of magnetic material (usually some form of steel). The flux is produced by a small magnetizing current which flows through the winding. The alternating magnetic flux induces an electromotive force (EMF) in the secondary winding magnetically linked with the same core and appears as a voltage across the terminals of this winding. Cold rolled grain oriented (CRGO) steel is used as the core material to provide a low reluctance, low loss flux path. The steel is in the form of varnished laminations to reduce eddy current flow and losses on account of this.

Typically, the coil connected to the source is known as the primary coil and the coil applied to the load is the secondary coil.

A schematic diagram of a single-phase transformer is shown in the Figure 1.1.

A single-phase transformer consists mainly of a magnetic core on which two windings, primary and secondary, are wound. The primary winding is supplied with an AC source of supply voltage V_1. The current $I\Sigma$ flowing in the primary winding produces flux, which varies with time. This flux links with both the windings and produces induced emfs. The emf produced in the primary winding is equal and opposite of the applied voltage (neglecting losses). The emf is also induced in the secondary winding due to this mutual flux. The magnitude of the induced emf depends on the ratio of the number of turns in the primary and the secondary windings of the transformer.

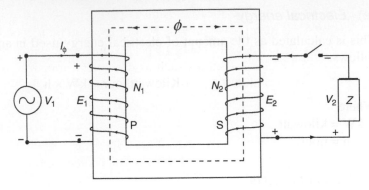

Figure 1.1
Schematic diagram of a single-phase transformer

Potential-induced

There is a very simple and straight relationship between the potential across the primary coil and the potential induced in the secondary coil.

The ratio of the primary potential to the secondary potential is the ratio of the number of turns in each and is represented as follows:

$$\frac{N_1}{N_2} = \frac{V_1}{V_2}$$

The concepts of step-up and step-down transformers function on similar relation.

A step-up transformer increases the output voltage by taking $N_2 > N_1$ and a step-down transformer decreases the output voltage by taking $N_2 > N_1$.

Current-induced

When the transformer is loaded, then the current is inversely proportional to the voltages and is represented as follows:

$$\frac{V_1}{V_2} = \frac{I_2}{I_1} = \frac{N_1}{N_2}$$

EMF equation of a transformer:
    rms value of the induced emf in the primary winding is:

$$E_1 = 4.44 \times f \times N_1 \times \phi_m$$

rms value of the induced emf in the secondary winding is:

$$E_2 = 4.44 \times f \times N_2 \times \phi_m$$

Where
    N_1 = Number of turns in primary
    N_2 = Number of turns in secondary
    ϕ_m = Maximum flux in core and
    f = Frequency of AC input in Hz.

1.1.3   Ideal transformer

The following assumptions are made in the case of an ideal transformer:

- No loss or gain of energy takes place.
- Winding has no ohm resistances.
- The flux produced is confined to the core of the transformer, which links fully both the windings, i.e., there is no flux leakage.
- Hence, there are no I^2R losses and core losses.
- The permeability of the core is high so that the magnetizing current required to produce the flux and to establish it in the core is negligible.
- Eddy current and hysteresis losses are negligible.

1.1.4   Types of transformers

1. As per the type of construction

    (a) *Core type*:   Windings surround a considerable part of the core.
    (b) *Shell type*:   Core surrounds a considerable portion of the windings.

2. As per cooling type

    (a) *Oil-filled self-cooled*:   Small- and medium-sized distribution transformers.
    (b) *Oil-filled water-cooled*:   High-voltage transmission line outdoor transformers.
    (c) *Air Cooled type*:   Used for low ratings and can be either of natural air circulation (AN) or forced circulation (AF) type.

3. As per application

    (a) *Power transformer*:   These are large transformers used to change voltage levels and current levels as per requirement. Power transformers are usually used in either a distribution or a transmission line.
    (b) *Potential transformer (PT)*:   These are precision voltage step-down transformers used along with low-range voltmeters to measure high voltages.
    (c) *Current transformer (CT)*:   These transformers are used for the measurement of current where the current-carrying conductor is treated as a primary transformer. This transformer isolates the instrument from high-voltage line, as well as steps down the current in a known ratio.
    (d) *Isolation transformer*:   These are used to isolate two different circuits without changing the voltage level or current level.

A few important points about transformers:

- Used to transfer energy from one AC circuit to another
- Frequency remains the same in both the circuits
- No ideal transformer exists
- Also used in metering applications (current transformer, i.e., CT, potential transformers, i.e., PT)
- Used for isolation of two different circuits (isolation transformers)
- Transformer power is expressed in VA (volt amperes)
- Transformer polarity is indicated by using dots. If primary and secondary windings have dots at the top and bottom positions or vice versa in diagram, then it means that the phases are in inverse relationship.

1.1.5   Connections of single-phase transformer

Depending on the application's requirement, two or more transformers have to be connected in a series or parallel circuits. Such connections can be undertaken as depicted in the following diagram examples:

(a) *Series connection of two single-phase transformers*

As shown in Figure 1.2, two transformers can be connected in a series connection. If both are connected as in Figure 1.2 then voltage twice that of voltage rating of the individual transformer can be applied. Their current rating must be equal and high enough to carry load current. Precaution should be taken to connect transformers windings, keeping in mind the polarity. In the above example, primary total turns to secondary total turns are in the 2:1 ratio, leading to half voltage.

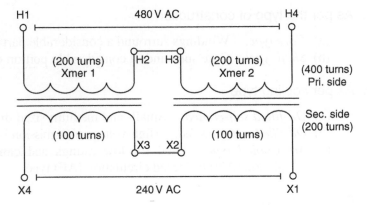

Figure 1.2
Series connection of two single-phase transformers

(b) *Parallel connection of two single-phase transformers*

As shown in Figure 1.3, two transformers are connected in series on the primary side while the secondary sides are connected in parallel.

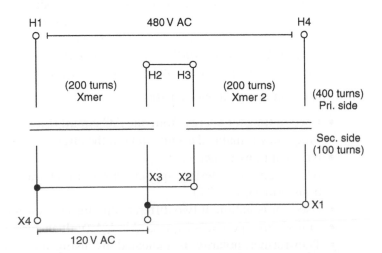

Figure 1.3
Parallel connection of two single-phase Transformers

On the primary side, the number of turns is added while on the secondary side they remain as it is due to their parallel condition.

LVDT (linear voltage differential transformer) is the best practical example of the basic transformer and its series connection. Use of transformers with such connections can pose problems of safety and load sharing and are hardly used in practical power circuits. It is possible to deploy these connections while designing control transformers if such use will have any specific advantage. Parallel operation of two separate transformers is possible under specific conditions to meet an increased load requirement but the risks involved must be properly evaluated.

1.1.6   Three-phase transformers

Large-scale generation of electric power is generally three-phasic with voltages in 11 or 32 kV. Such high three-phasic voltage transmission and distribution requires use of the three-phase step-up and step-down transformers.

Previously, it was common practice to use three single-phase transformers in place of a single three-phase transformer. However, the consequent evolution of the three-phase transformer proved space saving and economical as well.

Still, construction-wise a three-phase transformer is a combination of three single-phase transformers with three primary and three secondary windings mounted on a core having three legs.

Commonly used three-phases are:

- Three-phase three-wire (delta)
- Three-phase four-wire (star).

1. *Delta connection*

It consists of three-phase windings (Figure 1.4) connected end-to-end and are 120° apart from each other electrically. Generally, the delta three-wire system is used for an unbalanced load system. The three-phase voltages remain constant regardless of load imbalance.

$$V_L = V_{ph}$$

Where

V_L = line voltage
V_{ph} = phase voltage.

Relationship between line and phase currents:

$$I_L = \sqrt{3}\, I_{ph}$$

Where

I_L = line current
I_{ph} = phase current.

2. *Three-phase four-wire star connections*

The star type of construction (Figure 1.5) allows a minimum number of turns per phase (since phase voltage is $1/\sqrt{3}$ of line voltage) but the cross section of the conductor will have to be increased as the current is higher compared to a delta winding by a factor of $\sqrt{3}$. Each winding at one end is connected to a common end, like a neutral point – therefore, as a whole there are four wires.

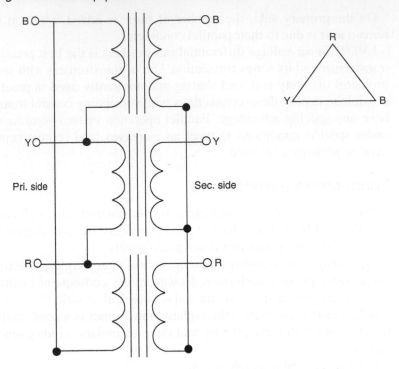

Figure 1.4
Three-phase transformer delta connection on primary side

A three-wire source as obtained from a delta winding may cause problems when feeding to a star connected unbalanced load. Because of the unbalance, the load neutral will shift and cause change of voltage in the individual phases of the load. It is better to use a star-connected four-wire source in such cases. Three-wire sources are best suited for balanced loads such as motors.

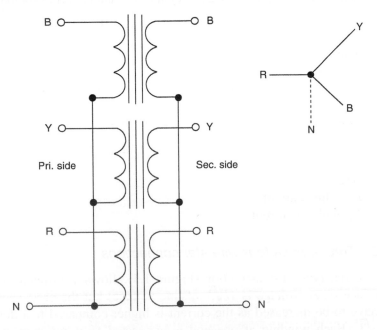

Figure 1.5
Three-phase four-wire transformer star connection

Relationship between line and phase voltages:

$$V_{L} = \sqrt{3} \ V_{ph}$$

Where

V_{L} = line voltage
V_{ph} = phase voltage.

Relationship between line and phase currents:

$$I_{L} = I_{ph}$$

Where

I_{L} = line current
I_{ph} = phase current.

Output power of a transformer in kW:

$$P = \sqrt{3} \times V_{L} \times I_{L} \times \cos\phi \ [\text{kW}]$$

Where

V_{L} = line voltage
I_{L} = line current
$\cos\phi$ = power factor.

3. *Possible combinations of star and delta*

The primary and secondary windings of three single-phase transformers or a three-phase transformer can be connected in the following ways:

- Primary in delta – secondary in delta
- Primary in delta – secondary in star
- Primary in star – secondary in star
- Primary in star – secondary in delta.

Figure 1.6 shows the various types of connections of three-phase transformers. On the primary side, V is the line voltage and I the line current. The secondary sideline voltages and currents are determined by considering the ratio of the number of turns per phase ($a = N_1/N_2$) and the type of connection. Table 1.1 gives a quick view of primary-line voltages and line currents and secondary-phase voltages and currents. The power delivered by the transformer in an ideal condition irrespective of the type of connection = 1.732 V_{L}, I_{L} assuming $\cos\phi = 1$.

1.1.7   Testing transformers

The following tests are carried out on transformers:

- Measurement of winding resistance
- Measurement of Voltage ratio
- Test phasor voltage relationship
- Measurement of impedance voltage, short-circuit impedance and load loss
- Measurement of no load loss and no load current
- Measurement of insulation resistance
- Dielectric test
- Temperature rise.

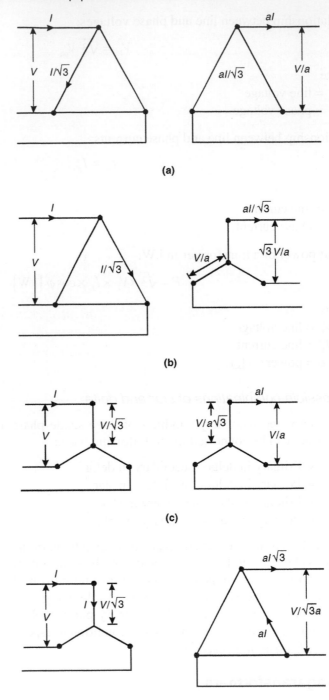

Figure 1.6
Types of connections for three-phase transformers: (a) Delta–delta connection; (b) Delta–star connection; (c) Star–star connection; (d) Star–delta connection

Connection	Line Voltage	Line Current	Phase Voltage	Phase Current
(a) Delta–delta				
Primary delta	V	I	V	$I/1.732$
Secondary delta	V/a	Ia	V/a	$Ia/1.732$
(b) Delta–star				
Primary delta	V	I	V	$I/1.732$
Secondary star	$1.732V/a$	$Ia/1.732$	V/a	$Ia/1.732$
(c) Star–star				
Primary star	V	I	$V/1.732$	I
Secondary star	V/a	Ia	$V/1.732\,a$	Ia
(d) Star–delta				
Primary star	V	I	$V/1.732$	I
Secondary delta	$V/1.732a$	$1.732\,Ia$	$V/1.732\,a$	Ia

Table 1.1
Voltage and current transformation for different three-phase transformer connections

Why is transformer rating defined in kVA?

A transformer, unlike a motor, has no mechanical output (expressed in kW). The current flowing through it can vary in power factor, from zero PF lead (pure capacitive load) to zero PF lag (pure inductive load) and is decided by the load connected to the secondary. The conductor of the winding is rated for a particular current beyond which it will exceed the temperature for which its insulation is rated irrespective of the load power factor.

Similarly, the voltage that can be applied to a transformer primary winding has a limit. Exceeding this rated value will cause magnetic saturation of the core leading to distorted output with higher iron losses.

It is therefore usual to express the rating of the transformer as a product of the rated voltage and the rated current (VA or kVA). This however does not mean that you can apply a lower voltage and pass a higher current through the transformer. The VA value is bounded individually by the rated voltage and rated current.

Why is power transmitted at higher voltages?

When a particular amount of power has to be transmitted over a certain distance the following aspects need to be considered to decide the best voltage.

A lower voltage the need higher size conductors to withstand the high current involved. There is a physical limitation to the size of conductor. Also, the percentage voltage drop may become excessive. A higher voltage will make the conductor size manageable and reduce the voltage drop (% value) but the cost of the line becomes high due to larger clearances needed.

The best voltage will be one in which the total operational cost which the sum of the annualized capital cost (of the line) and the running cost due to power loss in the line is the lowest. In practice, it is found that transmitting bulk power over long distances is economical if done in the HV range. The actual voltage will vary based on the distance and quantum of power. Distribution circuits where typically the amount of power and distance involved are both lower, the best voltage is in the MV range (11, 22 or 33 kV). For the same reason, low voltage circuits are found only in local sub-distribution circuits.

1.2   Basic principles of electrical machines

1.2.1   Electromechanical energy conversion

The electromechanical energy conversion device is a link between electrical and mechanical systems.

When the mechanical system delivers energy through the device to the electrical system, the device is called a generator.

When an electrical system delivers energy through the device to the mechanical system, the device is called a motor.

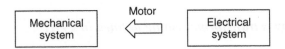

The process is reversible; however, the part of energy converted to heat is lost and is irreversible. An electric machine can be made to work either as a generator or as a motor. The electromechanical conversion depends on the interrelation between:

- Electric and magnetic fields
- Mechanical forces and motion.

In rotating machines, power is generated by the relative motion of the coils.

In the case of a generator, the winding is rotated mechanically in the magnetic field. This causes the flux linkages with the windings to change causing induced voltages.

In the case of a motor, the current-carrying conductor is allowed inside a magnetic field. Mechanical force is exerted on a current-carrying conductor in a magnetic field and hence a resultant torque is produced to act on the rotor.

In both a generator as well as a motor, the current-carrying conductor is in the magnetic field. The conductors and flux travel with respect to each other at a definite speed. In rotating machines, both voltage and torque are produced. Only the direction of power flow determines whether the machine is working as a generator or a motor. For a generator, e and i are in the same direction.

$$T_m = T_e + T_f$$

Where
T_m = mechanical torque
T_e = electrical torque
T_f = torque lost due to friction.

For a motor, e and i are in opposite direction.

$$T_m = T_e + T_f$$

In a generator, the power is supplied by the prime mover. Electrical power is produced by the action of the generator and the resultant power produced due to friction is lost. Whereas in the case of a motor, the power is supplied by the electrical power supply inputs, and there is a slight loss of the resultant mechanical power produced due to friction.

1.2.2   Basic principles of electromagnetism

Magnetic and electric fields

As you are aware, each electric charge has its own electric field; i.e., lines of force. Electric field lines point away from the positive charges and towards negative charges (Figure 1.7). Each charge exerts force on the other charge, which is always tangential to the lines of force created by the other charge.

Similarly, the magnetic field lines 'flow' away from the N-pole and towards the S-pole (Figure 1.8). A current moving the electric charges creates a magnetic field. Every orbiting electron forms a current loop that creates its own magnetic field. Magnetic field lines always form circles around the current creating them.

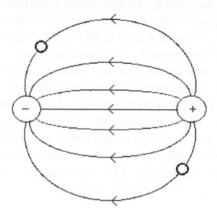

Figure 1.7
Electric force line of a charge

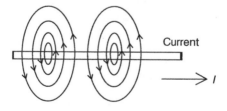

Figure 1.8
Magnetic field lines around a current-carrying conductor

Magnetic field produced by a current-carrying conductor

If a conductor carries a current, it produces a magnetic field surrounding it. The direction of the current and the direction of the field so produced have a definite relation that is given by the following rules:

The right hand rule   Hold the conductor in the right hand with the fingers closed around the conductor and the thumb pointing towards in the direction of the current. The fingers will point towards the direction of the magnetic lines of the flux produced around the conductor.

Flux produced by a current-carrying coil   Flux can be produced by causing the current to flow through a coil instead of a conductor. Introduction of magnetic material in the core on which the coil is wound increases flux. The direction of the magnetic flux in the coil is given by the right-hand rule.

In the case of a motor, the direction of the emf induced is such as to oppose the flow of current. Whereas, in a generator the emf induced is in such a direction as to establish a current.

Fleming's left hand rule   This defines the relationship between the direction of the current, the direction of field, and the direction of the motion. If the forefinger of the left hand points in the direction of the field, the middle finger points in the direction of the current, and the thumb points in the direction of the motion.

1.2.3   The basic principle of motor

The basic working of a motor is based on the fact that when 'a current carrying conductor is placed in a magnetic field, it experiences a force'.

If you take a simple DC motor, it has a current-carrying coil supported in between two permanent magnets (opposite pole facing) so that the coil can rotate freely inside. When the coil ends are connected to a DC source then the current will flow through it and it behaves like a bar magnet, as shown in Figure 1.9. As the current starts flowing, the magnetic flux lines of the coil will interact with the flux lines of the permanent magnet. This will cause a movement of the coil (Figures 1.9(a), (b), (c), (d)) due to the force of attraction and repulsion between two fields. The coil will rotate until it achieves the 180° position, because now the opposite poles will be in front of each other (Figure 1.9(e)) and the force of attraction or repulsion will not exist.

The role of the commutator: The commutator brushes just reverse the polarity of DC supply connected to the coil. This will cause a change in the direction of the current of the magnetic field and start rotating the coil by another 180° (Figure 1.9(f)).

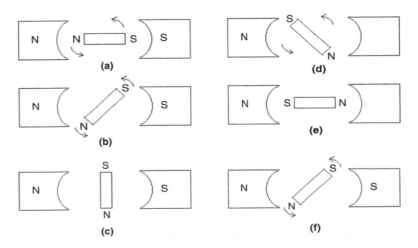

Figure 1.9
A motor action

The brushes will move on like this to achieve continuous coil rotation of the motor.

Similarly, the AC motor also functions on the above principle; except here, the commutator contacts remain stationary, because AC current direction continually changes during each half-cycle (every 180°).

1.2.4   Basic principle of generator

We have discussed the basic working of a motor and through the diagrams we have seen a generator action as well.

In principle, an AC generator's construction is similar to the construction of the motor. Instead of putting current in, current is taken out from the coil in an alternator.

A mechanical prime mover rotates the coil in between the poles of a permanent magnet and an AC potential is induced in the coil. To further define: if an AC current will make a coil turn, then turning the coil will create an AC current.

As per Faraday's law, when a wire is moved in to cut across magnetic field lines, a force is exerted on the charge (electrons) in the wire by trying to move them along the wire. This is how current will start flowing if a complete circuit is provided to it. The magnetic field is provided not by magnets, but by field coils.

The coil in which the voltage is induced is called armature winding, while the coil that provides the magnetic field is called field winding.

In high-voltage generators, it is not good practice to have armatures rotating because current-collecting brushes of high ratings are required. Rather, the armature is kept stationary and the field is kept rotating.

Alternators of low capacity use a permanent magnet as a field, while in high-capacity alternators field winding supply is derived from the exciter assembly. An exciter assembly is a small alternator connected on the same shaft.

1.2.5   Idealized machines

There is a stationary member called a stator and a rotating member called a rotor. The rotating member is mounted on bearings fixed to the stationary member. The stator and the rotor have cylindrical iron cores, separated by an air gap. Windings are wound on the stator and the rotor core. A common magnetic flux passes across the air gap from one core to another forming a combined magnetic circuit. Two cylindrical iron surfaces with an air gap between them move relative to each other. The cylindrical surface may be divided by an even number of salient poles with spaces in between, or it may be continuous with slot openings uniformly spaced around the circle. This structure may be for either of the stator or the rotor.

The common features of an ideal electrical machine are shown in the Figure 1.10. For windings, conductors run parallel to the axis of the cylinders near the surface. The conductors are connected into coils by the end connections outside the core and the coils are connected to form the windings of the machine.

The operation of the machine depends on the distribution of the currents around the core surfaces and the voltages applied to the windings.

In various types of electrical machines, the arrangement differs in the distribution of the conductors, windings, and in core constructions, depending on whether it is a continuous or a salient pole type. The magnetic flux permeates the iron cores in a complex manner. However, as the iron has a high permeability, the accurate working of a machine can be determined by considering the flux distribution in the air gap. The conductors are actually located in slots formed in the laminations of the core.

A typical cross section and the corresponding development diagram of an electrical machine with four poles, perpendicular to the axis of the cores is shown in Figure 1.11. As shown in the diagram, the distribution of flux and current repeats itself at every pair of poles. On the poles, the windings are so wound that the current flows in the opposite direction and produces a field corresponding to the north and south polarities. Maximum flux is along the center of the pole and reduces to zero between the interpole gaps.

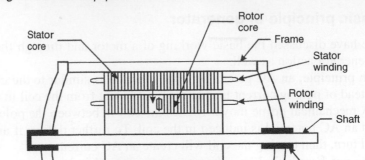

Figure 1.10
Common features of an ideal electric machine

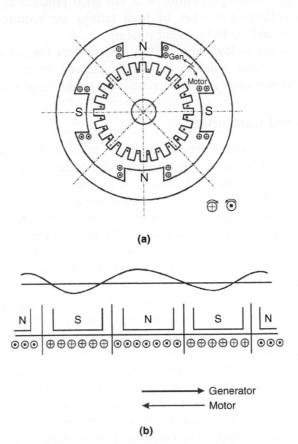

(a)

(b)

Figure 1.11
Typical cross section and development of an electrical machine

1.2.6 Basic principles of electrical machines

In an electrical machine, the currents in all the windings combine to produce the resultant flux. The field system produces flux. Voltages are induced in the windings such as those of an armature. When the armature carries current, the interaction between the flux and the current produces torque.

Types of electrical machine windings

(a) *Coil winding*

The winding consists of coils wound on all the poles of the machine and connected together to form a suitable series or parallel circuit. The direction of the current in the alternate pole will be opposite so that when one pole is the North Pole, the other adjacent pole will be a South Pole. This produces the flux in the proper direction, completing the magnetic circuit from the North Pole to the South Pole through the iron cores of both the stator and the rotor.

The coil may be wound on the stator or on the rotor, forming the salient or non-salient poles of the machine. The DC supply is given to these windings and they produce a field proportional to the magnitude of the current through the windings. If the poles are on the stator, a stationary field is produced in the air gap.

(b) *Commutator winding*

The commutator winding is on the rotor. The armature has open slots and the conductors are located in these slots and connected to the commutator segments in a continuous sequence.

(c) *Polyphase winding*

Polyphase winding is a distributed winding. Individual conductors are distributed in slots in a suitable way and connected into a number of separate circuits, one for each phase. The group of conductors forming the phase bands is distributed in a regular sequence over the successive pole pitches so that there is balanced winding that produces an equal voltage per phase. This type of winding is mainly used for the stator. When supplied with three-phase currents it produces a rotating field in the air gap. This is of a constant magnitude but rotating at a constant synchronous speed.

1.2.7   Types of electrical machines

Depending on the type of combinations of windings used on the stator and the rotor, electrical machines are classified in different types as follows:

1. *DC machines*

The DC machines have an edge over AC machines when it comes to the speed control of a motor. It is easier and cheaper.

(a) *Shunt motor*

This machine has field winding mounted in yoke and the armature winding is mounted on rotor. The shunt motor is used where speed regulation is important.

Self excited   Field winding is connected in parallel (shunt) with the armature winding on the same supply. Changing the field current can vary the speed. Torque is proportional to armature current.

This machine can also act as a generator. To limit the high starting current of the motor drive release the voltage in the ramp. For this motor, a variable resistor is connected in series with the field circuit to change the flux value and the speed by a small amount.

Separately excited   Field winding is connected in parallel (shunt) with the armature winding with separate excitation.

Torque is proportional to armature current. In a separately excited shunt motor, speed can be varied up to a certain limit by changing armature voltage. After that using field weakening (reducing field current), it is possible to increase the speed of motor above base speed. Other features remain same as that of the self-excited one.

(b) *Series motor*

As the name suggests in this type of motors, field winding is connected in series with the armature winding. Naturally, heavy current will pass through it; hence field winding of a thicker gage is used. A series motor is used where speed regulation is not important.

The main advantage of this motor is that a high torque can be obtained, which makes it useful for applications such as diesel locomotives, cranes, etc.

The relationship between Torque and current is as follows:

$$T \, \alpha \, Ia^2$$

It is important to start this motor in a loaded condition else it could lead to damage of the motor and its surroundings.

(c) *Compound motor*

If we combine both series and shunt motors then we will have a compound motor. This combines the good features of both types such as high torque characteristics of a series motor and the speed regulation of a shunt motor.

2. *AC machines*

(a) *Squirrel-cage induction motor*

AC machines are simple and sturdy. The most common machine of this type is the Squirrel cage induction motor (the name was derived from its construction type). The basic working of this was dealt with in the previous sections. The following relation gives the speed of this motor:

$$N \ (\text{rpm}) = \frac{120 \, f}{p}$$

Where
  f = frequency
  p = number of poles.

For example, if the motor has two poles, then at 50 Hz frequency the motor rpm will be 3000 (rpm).

However, you will not find 3000 or 1500 rpm on the motor nameplate because the motor rpm will not be 3000 rpm at full load. This is because of a slip associated with an induction motor.

The RPM of the motor is controlled by controlling the frequency (f) – as frequency increases, motor speed will also increase. High starting current is limited using a star/delta starter or reduced-voltage starters.

(b) *Wound rotor motor*

This is similar in construction to the squirrel cage and works similarly too, except that slip rings are provided. The main feature of the slip ring motor is that resistors, which are connected in series with the rotor circuit, limit the starting current.

The motor starts with a full resistance bank, but as speed of the motor increases, the resistances are shorted, one by one. As the motor reaches full speed, the whole bank of resistance is shorted out and the motor now runs as a normal induction motor.

(c) *Synchronous motors*

Synchronous motor is a constant-speed motor, which can be used to correct the power factor of the three-phase system. Like the induction motor in terms of the stator, the synchronous machine has either a permanent magnet arrangement or an electromagnet (with current supplied via slip rings) rotor. In simple terms, the rotor will keep locking with the rotating magnetic field in the stator. So, a two-pole machine will run at exactly 3000 rpm. In many synchronous machines, a squirrel cage is incorporated into the rotor for starting. Therefore, the machine acts as an induction motor when starting and as it approaches synchronous speed, it will suddenly 'lock in' to the synchronous speed.

1.2.8 Basic characteristics of electrical machines

The following are the basic characteristics of electrical machines:

- The voltages induced in the windings, the load currents, and the terminal voltages depend on the following different loading conditions – the speed at which the machine works under different loading conditions and the frequency.
- The power input or the output received from the machine.
- The torque produced under different loading conditions.

1.3 AC power systems

1.3.1 Single-phase power system

The power in a single-phase system is shown in Figure 1.12. In the figure current (I) lags voltage (V) by an angle ϕ. The current has two components – the energy component and the watt less component. Only the energy component has a power value. Hence, the power in a single-phase circuit is given by the following equation:

$$V \times I \times \cos \phi$$

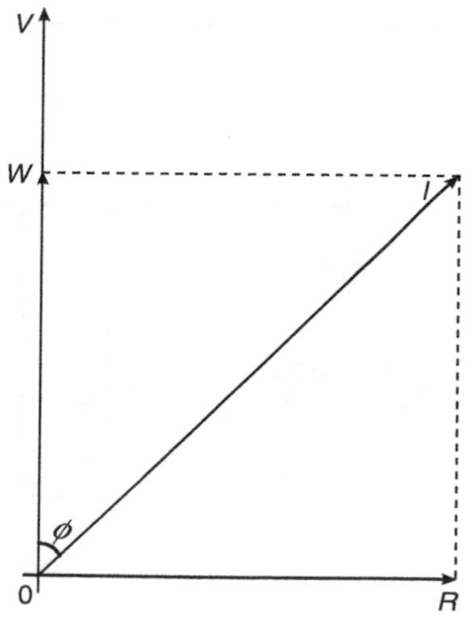

Figure 1.12
Single-phase power system

Where

P = power (watts)
V = voltage (rms)
I = current (rms)
$\cos \phi$ = power factor.

$$\frac{P}{V \times I}$$

1.3.2   Three-phase power systems

The three windings of a three-phase transformer or an alternator can be connected in either of two ways – delta or star as shown in the Figure 1.13. The relationship between the phase voltages and the currents, and the line voltages and the currents are as follows:

- *For delta-connected system*

  Line voltage = Phase voltage
  Line current = 1.732 × Phase current

- *For star-connected system*

  Line voltage = 1.732 × Phase voltage
  Line current = Phase current

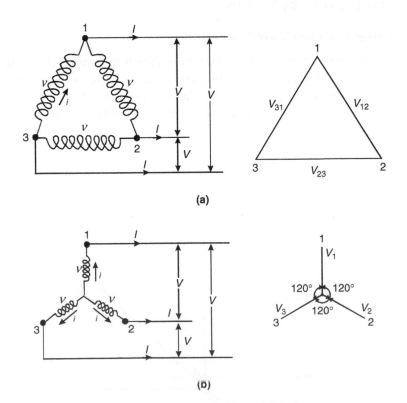

Figure 1.13
Three-phase circuit connections: (a) Three-phase delta connection; (b) Three-phase star connection

In a star connection, a neutral point is available. Generators are generally star-wound and the neutral point is used for earthing. Three-phase motors can be either delta- or star-connected. Usually, delta connections are used for low-voltage, small-size motors to reduce the size of the windings.

Three-phase currents are determined considering each phase separately and calculating the phase currents from the phase voltages and impedances. In practice, the calculations are simple and straightforward as three-phase systems are usually symmetrical, the loads being balanced. Most of the three-phase motors can be assumed as balanced loads. The calculations for currents, power, etc. can be done using the expression given below. However, for unsymmetrical or unbalanced systems, the calculations and expressions given below do not hold good, and complex calculations are required.

The power in a three-phase system is the sum of the power of the three phases. Let us consider a balanced delta- or star-connected system. The total power for the three-phase system will be:

$$P = 1.732 \times V \times I \times \cos\phi$$

Where
V = Line voltage
I = Line current.

1.3.3   Power measurement in a three-phase system

Electrical power is measured with a wattmeter. A wattmeter consists of a current coil connected in series with load, while the other potential coil is connected parallel with load. Depending on the strength of each magnetic field movement, the pointer gets affected.

The true or real power is directly shown in a wattmeter. In three-phase systems, power can be measured using several methods. For temporary measurements, a single wattmeter can be used. However, for permanent measurements, a three-phase wattmeter having two elements is used which indicates both balanced and unbalanced loads. For an unbalanced load, two wattmeters must be used as shown in the Figure 1.14. The total power is calculated by adding the measurement readings given by the two wattmeters. With this method, the power factor can also be obtained.

When using the two-wattmeter method, it is important to note that the reading of one wattmeter should be reversed if the power factor of the system is less than 0.5. In such a case, the leads of one wattmeter may have to be reversed in order to get a positive reading. In the case of a power factor less than 0.5, the readings must be subtracted instead of being added. The power factor of the three-phase system, using the two-wattmeter method (W_1 and W_2) can be calculated as follows:

$$\tan\phi = \frac{1.732\left(W_1 \times W_2\right)}{\left(W_1 + W_2\right)}$$

Since the sum and subtraction of readings are done to calculate total true power of a three-phase system, methods shown are not used practically in industry. Rather three-phase power analyzers are used which are more user-friendly.

Power factor meter

It is similar to a wattmeter in principle, only two armature coils are provided with mountings, on a single shaft. They are 90° apart from each other. Both armature coils rotate as per their magnetic strengths. One coil moves proportional to the restive component of the power, while the other coil moves proportional to the inductive component of the power.

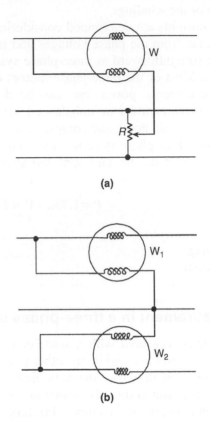

(a)

(b)

Figure 1.14
Methods of measuring the power in three-phase systems: (a) One wattmeter method for balanced load; (b) Two wattmeter method for balanced/unbalanced loads

Energy meter

This shows the amount of power (electric energy) used over a certain period. In a watt-hour meter, there are two sets of windings. One is the voltage winding while the other is the current winding. The field developed in the voltage windings causes current to be induced in an aluminum disk. The torque produced is proportional to the voltage and current in the system. The disk in turn is connected to numeric registers that show electric energy used in terms of kilowatt-hours.

1.4   Meters used in troubleshooting

For troubleshooting electrical circuits and systems, the following meters are used depending on the requirements of the parameters to be measured or detected for faultfinding:

- Multi-range voltmeters
- Clip-around or clamp-on ammeters
- Electrostatic voltmeters (high-voltage measurements)
- Multimeter or volt-ohm-milli-ammeter (voltage, resistance, current, etc.)
- Thermocouple meters (indirect current measurement)

- True wattmeters (directly measurement of power in watts)
- Pseudo wattmeters
- Digital voltmeters
- Heterodyne wavemeter (analog measurement of frequency)
- Digital frequency meters
- Continuity testers
- Analog ohmmeters
- Digital ohmmeters
- Insulation testers
- Digital capacitance meters
- Q meter (measuring inductance and capacitance)
- Oscilloscope (measuring wave forms, amplitude, frequency, phase, etc.)
- Dip meter (radio frequencies)
- Logic level probe testers
- Logic analyzers (diagnosing logic systems problems)
- Spectrum analyzers.

Certain important meters from the above list are explained in detail in the following chapters as and when required.

2

Devices, symbols, and circuits

Objectives

- To understand basic electrical symbols
- To understand power and control circuits
- To read electrical drawings.

2.1   Devices and symbols

Any electrical drawing representing an electrical installation or a circuit takes the help of specific symbols to represent various electrical devices in shorthand. This provides a quick idea to the reader about a circuit or installation, and is particularly useful while troubleshooting.

Therefore, it is important to familiarize oneself with various symbols. Some of the commonly used device symbols are detailed in the following section and in Figure 2.1.

2.2   Electrical circuits

Electrical circuits are circuits used to interconnect different electrical equipments together to enable the working of an electrical device.

Electrical schematics are commonly classified into power circuit and control circuit. A power circuit consists of the main power device (a motor, a generator, or other power devices) along with heavy power conductors, contactors, protection devices.

A control circuit consists of switches, field device contacts, timers, relay coils, relay contacts, protection devices, and light power conductors.

2.2.1   Power circuits

Power circuits are required for carrying power to or from heavy electrical equipments like motors, alternators, or any electrical installation.

They carry out the following functions:

- Isolation using devices such as isolators, linked switches and circuit breaks.
- Circuit control using devices such as contactors, motor circuit breakers, etc.
- Protection against overload and short-circuits using thermal overload relays, electro-magnetic relays, circuit breakers, with releases, fuses, etc.

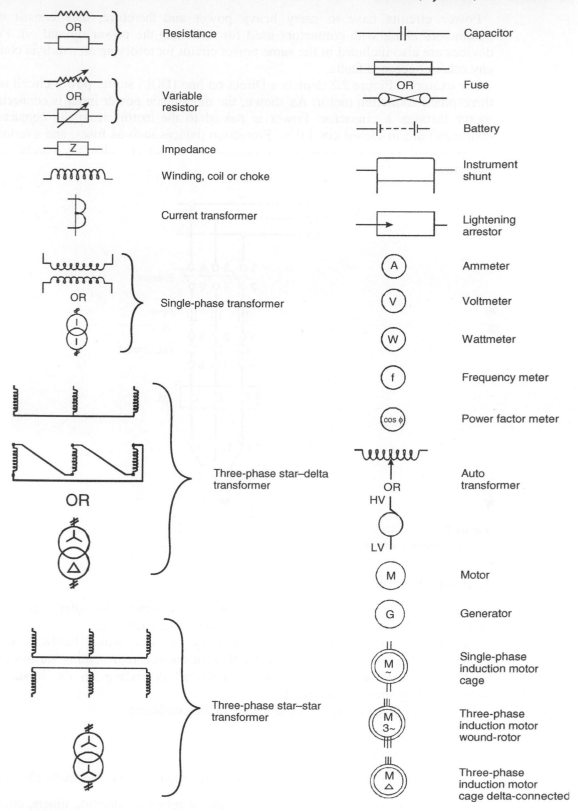

Figure 2.1
Electrical devices and symbols

Power circuits have to carry heavy power and therefore, they consist of heavy conductors along with contactors used for switching the power on and off. Protection devices are also included in the same power circuit for resolving an overload condition or any other concerned faults.

For example, Figure 2.2 depicts a Direct-on-line (DOL) starter power circuit used for a three-phase induction motor. As shown, the three-phase power input is connected to the motor through a contactor. Power is passed to the motor when the contacts (of the contactor) are in closed condition. Protection devices such as fuses, and overload relays are provided in series with power conductors to detect unhealthy conditions during operation.

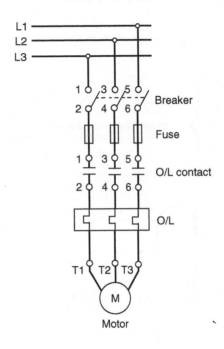

Figure 2.2
Power circuit for a motor

2.2.2   Control circuit

A control circuit is for the automatic control of equipment, for safety interlocking, and sequencing the operations of the plant equipment and machines.

Control circuits hardware consists of relay contacts, wires, hardware timers, and counters, relay coils, etc. These consist of input contacts representing various conditions; the output coils are energized or de-energized depending on the input conditions represented by the control circuit.

Input contacts represent the binary state of the condition:

- True or false
- On or off.

There are two types of contacts NO (normally open) and NC (normally closed).

- *Input contact*:   These are contacts of relays, contactors, timers, counter, field instrument switches, pressure switches, limit switches, etc.
- *Output coil*:   These have two states – On or Off. Output coil can be auxiliary contactor or Main contactor coil.

A few simple control circuits are shown in Figure 2.3 to represent logical AND, OR, and such conditions.

1. 'AND' operation circuit

Figure 2.3(a) shows a simple control circuit (AND operation) with two input contacts (NO) representing two conditions that must be true to complete the circuit to switch on the output relay coil and change the state of output from 'Off' to 'On'.

2. 'OR' operation circuit

Figure 2.3(b) shows a circuit with three input contacts (NO) representing that at least one of the three conditions should be true to complete the circuit to switch On the relay coil and change output state from 'Off' to 'On'.

3. 'AND with OR' operation circuit

Figure 2.3(c) shows a control circuit, consisting of a combination of AND and OR operations.

There are two parallel (OR condition) paths with two input contacts (NO) connected in series in each path representing AND conditions. The path for coil K3 will be completed when one of the path conditions comes true. The circuit then will switch 'On' the relay coil and change the output state from 'Off' to 'On'.

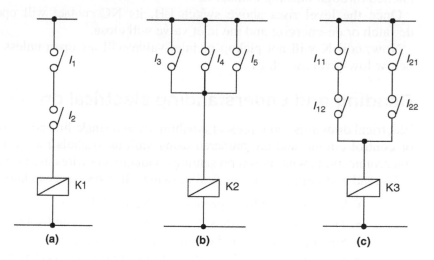

Figure 2.3
Simple control circuits

Example 2.1

Design control circuit for 'Tank water level control'.
Operation sequence should be such that

- When the water level goes below the low-level limit, open the inlet valve of water tank.
- When water level goes above high limit is detected, close the inlet valve.

Build a control circuit for the same.

As shown in Figure 2.4, when the level is initially low, coil K will pickup (since both level switch NC contacts will remain as it is), thus energizing the inlet valve to open.

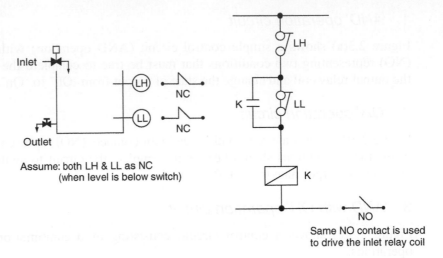

Figure 2.4
Example of simple control circuit for water tank inlet valve operation

As the level rises above switch LL, its NC contact will open but still coil K will remain latched through latching contact K.

Once the level rises above switch LH, its NC contact will open causing coil K to de-latch or de-energize and the inlet valve will close.

Now, coil K will not pickup or inlet valve will not open unless the water level drops below low-level switch LL.

2.3   Reading and understanding electrical drawings

Electrical drawings can represent anything from a single-line power distribution, to a power or control circuit, and are prepared using various symbols for electrical devices and their interconnections with lines representing conductors or wires used for interconnections.

To read and understand electrical drawings, it is necessary to know the following:

- Symbols used for representing electrical devices
- Their interconnections, legends, terminology, and abbreviations
- Sheet numbering and column format for each sheet
- Wire and terminal numbering (an important aspect in understanding electrical drawings).

Wire and terminal numbers are quite useful during installation and termination of cables, and during fault finding and troubleshooting.

It is easy to trace the connections and continuity of wires, if the wires and terminals are numbered using thorough cross-referencing terminology. Various examples of electrical schematics are shown in this section to explain the drawing methodology in practical circuits and in the interest of simplifying the scheme the following have not been included. These are however a MUST and will be insisted upon by the regulatory agencies.

- Any power circuit has to be provided with an isolating mechanism which usually includes the fuses also in the form of a switch-fuse unit. The schematics here depict only the fuse.

An emergency switch or a push button has to be provided near a mechanism to positively isolate the electrical circuit feeding the mechanism in the case of any emergency/accident. The NC contact of such a push button is connected in series with the other control contracts such as overload relay. The push button mechanisms are lockable and need a key to release once the push button is pressed.

2.3.1 Things to look for in an electrical drawing

1. The symbols shown for a device in a circuit represent its de-energized state when no power is applied. It is either a timer NO/NC contact or a relay NO/NC contact in a circuit. In addition, power devices such as circuit breakers and contactors are provided with NO and NC auxiliary contacts which are used for indicating the status of the device in signaling and interlocking circuits.
2. An electrical drawing has a sheet number and each sheet is divided into columns listed vertically as A, B, C, D and horizontally as 1, 2, 3, 4. This kind of matrix arrangement helps in quickly locating a particular device or contact in a sheet. Similarly, it is used to cross-reference a contact.
3. In order to identify different coils and their contacts a letter such as K1, K2 or C1, C2 is placed next to the circle of the coil. Contacts of the same contactor coil are shown with the same letter in the drawing.
4. Particular relay contacts may be used in different circuits at different locations. To give the reader an exact idea of where it is used, a drawing mentions a cross-reference number for each contact showing the sheet number and its matrix number.
5. In general, a heavy line is used to show high current-carrying conductors (mains supply lines, motor connection leads). In contrast, light-looking lines are used to represent low current-carrying conductors (control circuit lines).
6. Control circuit power lines are shown as L1 and L2; load (coils of relay) is connected between these two lines in series with switches, fuses, etc.
7. Conductors intersecting each other with no electrical junction in between are represented with an intersection without any dot. Conversely, conductors having an electrical junction are represented with a dot at the intersection.
8. A broken line in an electrical circuit represents mechanical action. Generally, it is associated with a push button or switch closing or opening a contact. Sometimes these lines can also indicate in combination with suitable additional symbols, a mechanical interlocking between two or more devices such as contactors or circuit breakers.
9. Dotted lines are used to differentiate an enclosure from field devices.
10. A wiring diagram of electric equipment represents the physical location of the various devices and their interconnections.
11. In an electrical drawing, conductors are marked with cross lines and dimensions of conductors are given alongside. This is used to represent the conductor size of a particular section in a drawing.

Based on the above hints, let us consider a few common examples of electrical drawings.

Example 2.2

Three-phase motor with DOL starter

This is depicted in an electrical drawing in Figure 2.5 along with the power and control circuits.

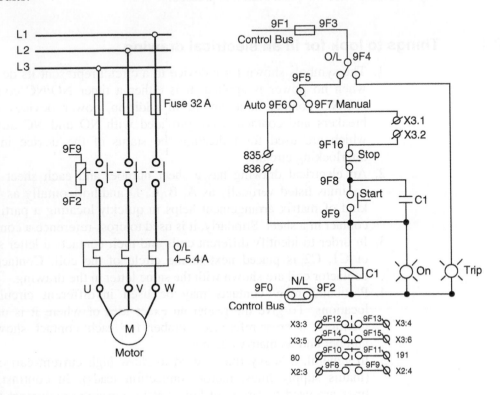

Figure 2.5
Typical electrical drawing of power and control circuits for a three-phase motor with DOL starter

The power circuit consists of a three-phase main supply with a fuse unit for protection purposes. The other side of the fuse unit is connected to a power contactor. The output terminals of the contactor are connected to an overload relay. Finally, the overload relay output terminals are connected to motor terminals.

The control circuit for the motor works on a 110 V AC single-phase supply. The phase of the control supply is connected to a NC contact of the overload relay (O/L). The wire from the O/L relay contact is connected to an auto/manual mode selector switch.

In the auto mode, the motor gets a start/run command through a potential-free contact (Terminals 835–836) of a relay, which in turn is energized with a Programmable Logic Controller (PLC) output.

In the manual mode, the motor can be started with the help of a start pushbutton. When the start pushbutton is pressed, the control circuit is completed and the auxiliary control contactor (C1) coil is energized. A potential-free NO contact of the contactor (C1) is closed and keeps the contactor C1 latched when the start pushbutton is released. When the auxiliary contactor (C1) is on, the motor power circuit is completed and the motor starts and remains on until the contactor C1 is de-energized and the power circuit to the motor terminals is broken.

For the manual mode, additional interlocks to trip the motor are connected between terminals X3.1 and X3.2. The motor can be stopped with a stop pushbutton. The NC contact of a stop pushbutton breaks the control supply to the auxiliary control contactor (C1) and the motor is stopped. The neutral for the control circuit is connected with a neutral link (N/L).

To indicate that the motor is ON or running, an indication lamp is connected in parallel to the contactor, which goes ON whenever the auxiliary contactor is turned on.

Another indication lamp to indicate a motor trip is connected to a NO contact of the overload relay. When the motor is overloaded, the NO contact is closed and the TRIP indication lamp is turned ON, until the overload relay is reset.

In the control circuit, potential-free contacts, 2 NO and 2 NC, of the auxiliary contactor (C1) are connected to various pairs of terminals such as X3:3 – X3:4 (NC), X3:5 – X3:6 (NC), 80 – 191 (NO), and X2:3 – X2:4 (NO).

The NO contact of the auxiliary control contactor is terminated at the terminals X2:3 and X2:4, and is used in parallel to the start pushbutton NO contact for latching purposes.

In addition, the contact letters 9F8-9F9 are mentioned. This shows the location of the contact in the drawing.

Example 2.3

Three-phase motor with star–delta starter

The electrical drawing in Figure 2.6 depicts this power circuit.

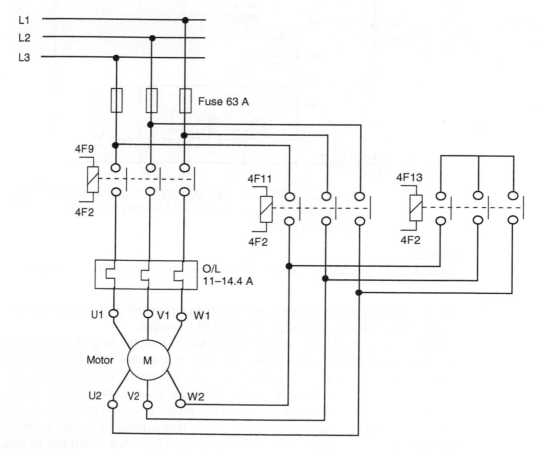

Figure 2.6
Typical electrical drawing of power circuit for a three-phase motor with star–delta starter

The power circuit consists of a three-phase mains supply with a fuse unit, three contactors – line contactor, star contactor, and delta contactor. The line contactor gets its three-phase power supply from the fuse unit and the output terminals of the line contactor are connected to the overload relay.

Overload relay output terminals are connected to the motor terminals – U1, V1, W1. Motor terminals U2, V2, W2 are connected through either star or delta contactors.

The star contactor and delta contactor are mutually interlocked in the control circuit to ensure only one contactor is on at a time. When the delta timer is on, the motor-winding terminals – U2, V2, W2 – get a three-phase supply and the motor is delta-connected. When the star contactor is on, the motor terminals – U1, V1, W1 – are shorted and the motor is star-connected.

The control circuit as shown in Figure 2.7, for the motor, works on a 110 V AC single-phase supply. The phase of the control supply is connected to a NC contact of the overload relay (O/L). The wire from the O/L relay contact is connected to an auto/manual mode selector switch.

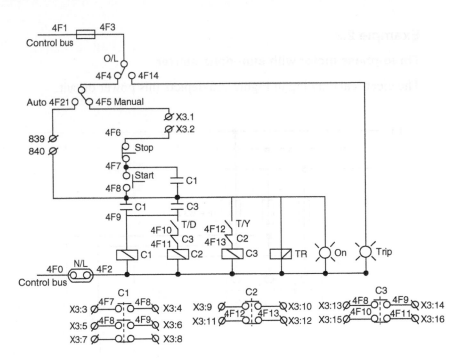

Figure 2.7
Typical electrical drawing of control circuit for a three-phase motor with star–delta starter

In an auto mode, the motor gets a start/run command through a potential-free contact of a relay, which in turn is energized with a PLC output.

In the manual mode, the motor can be started with the help of a start pushbutton. When the start pushbutton is pressed momentarily, the control circuit is completed and the line contactor is energized. A potential-free NO contact of the line contactor is closed and keeps the control complete when the start pushbutton is released. When the motor is started, the star contactor is closed and the motor is started with a star connection. As the motor runs for a few seconds, the delta timer picks up, which energizes the delta contactor and de-energizes the star contactor. The motor continues to run, connected in the delta configuration, until it is stopped with the stop pushbutton or trips due to an overload or an external interlock.

As can be viewed in Figure 2.7, each contactor used contacts that are given at the end of the drawing. For example, NO contact used is shown with letters 4F7-4F8 and 4F8-4F9 specify their locations in the drawing. Similarly, contact details of contactor C2 and C3 are shown.

Note: The overload relay in this circuit is actually connected in series with the phase winding of the motor in the normal running mode (i.e., delta connection). The motor rated current is normally indicated in terms of the line current which is greater than the phase current by a factor of $\sqrt{3}$. Selection and setting of the overload relay must take this into consideration.

Example 2.4

Let us consider electrical drawings of an inverter drive, as shown in Figures 2.8 and 2.9.

Figure 2.8 shows the power circuit wiring for the motor and control circuit wiring for starting and stopping the motor. The three-phase power supply is passed through the fuses and a contactor (1K1) and is connected to an incoming choke (CH1). The output of the choke (CH1) is connected to the input terminals of the inverter drive. The inverter drive gets its main power only when the contactor (1K1) is on. The inverter drive output supply is connected to an output choke (CH2) and the output of the choke (CH2) is connected to the three-phase motor terminals. The inverter drive and the motor are earthed.

The control circuit for the inverter drive works on a 110 V AC single-phase supply. The control circuit for contactor (1K11) consists of the following series of potential-free contacts:

1. Drive OK (NO contact of 1K12)
2. Emergency stop (NO contact of 1K13)
3. Local stop pushbutton (NC contact)
4. Remote stop pushbutton (NC contact)
5. Local/remote selector switch changeover contacts
6. Start pushbutton (NO contact).

The contactor 1K11 is energized when the control circuit is completed.

The contactor 1K1 is energized, when the drive output contact is closed and additional interlocks connected between the terminals 1X11:11 and 1X11:12 are OK.

A NO contact (13–14) of 1K1 is used to turn on the indication lamp (L2) to indicate that the drive is on. An NC contact of 1K1 is used for indicating a drive trip by turning the lamp (L3) on.

Another contactor (1K12) is energized to indicate drive OK, using a 24 V DC supply through a potential-free contact drive O/P (Terminal X100: 6–7).

Figure 2.9 shows the wiring diagram for the inverter drive control terminals. The inverter drive has the following sets of terminals:

• X100: Contacts for drive OK status
• X101: For start/stop (13–16), reset faults (13–18) commands to the inverter drive
• X102: For remote speed reference (25–27–28) for the inverter drive and analog outputs for speed indication (34–35)
• X9: Main contactor ON (4–5) and power supply (1–2) for external use.

As shown in the figure, terminal grouping is based on different operational functions. For example, digital contacts of drive are grouped with letter X101; whereas analog speed reference input and rpm display output are grouped with letter X102.

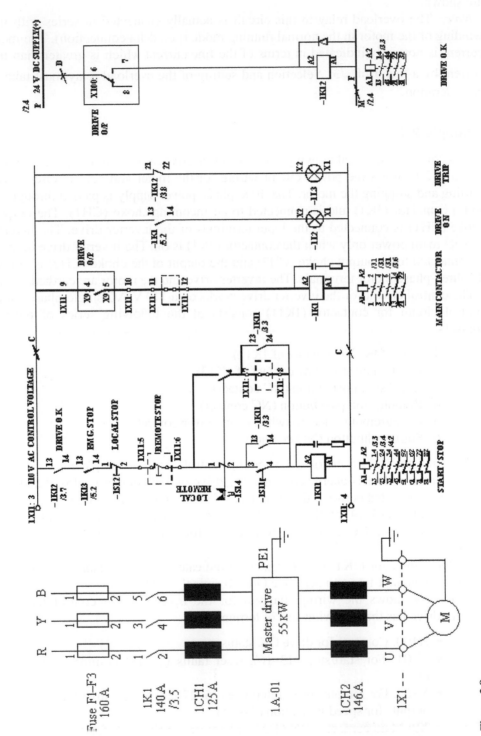

Figure 2.8
Power and control circuit for an inverter drive

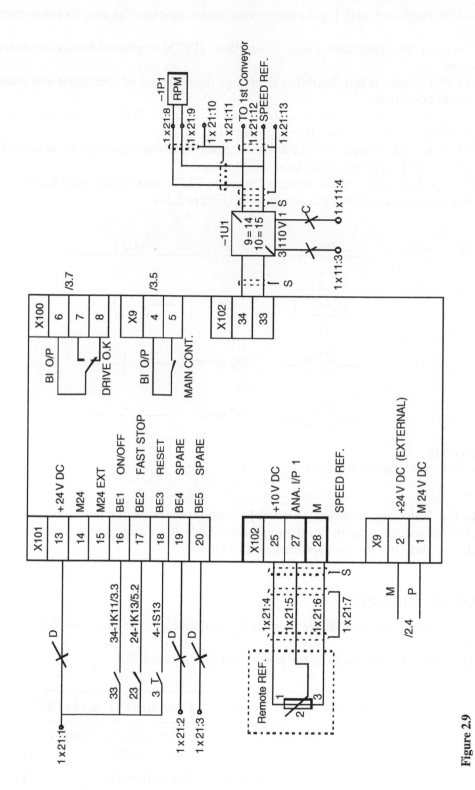

Figure 2.9

Control circuit with wiring terminals for an inverter drive

2.4   Reading and understanding ladder logic

Once the hardwire relay logic concepts are understood then its easy to comprehend ladder logic.

The term 'Programmable Logic Controllers' (PLCs) originated from relay-based control systems.

In a PLC, there is full flexibility to change the sequence of operations and interlocks for different conditions.

There are integrated circuits and internal logic in the PLC, in place of discrete relays, coils, timers, counters, and other such devices.

PLCs provide greater computational capabilities and accuracy, to achieve increased flexibility and reliability, than hard-wired relays.

The symbols and control concepts used in PLCs come from relay-based control and form the basis of ladder logic programing (Figure 2.10).

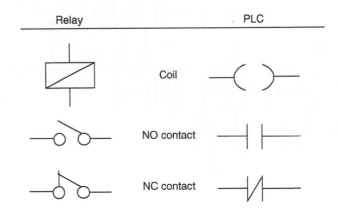

Figure 2.10
Comparison of relay and PLC terms

In the following sections, commonly used terminology for ladder logic is dealt with. The terminology used in commercially available PLCs from various manufacturers may differ slightly but the concepts remain the same.

2.4.1   PLC terminology

PLC terminology may differ from relay terminology, but the control concepts are the same.

The following are some of the terms used in relays and PLC:

Terms Used for Relay	Equivalent Terms in PLC
Contact input or condition	Coil output or temporary working bit
NO contact of relay condition	Normally open
NC contact of relay condition	Normally closed

As such, there is no equivalence between these terms. The term 'condition' is only used to describe ladder logic diagram programs in general and is equivalent to a set of basic instructions. The terms input/output are used for reference to I/O bits assigned to input and output signals.

In ladder logic programing, the following two types of instructions are used:

1. Instructions that correspond to the conditions of the ladder logic diagram. They are used in instruction form only when converting a program to mnemonic code.
2. Instructions that are used on the right-hand side of the ladder logic diagram are executed according to the conditions on the instruction lines preceding them.

Most of the instructions have at least one or more operands.

2.4.2 Ladder logic diagram

A ladder logic diagram is so-called because the relay logic runs in parallel lines between two power lines and the whole diagram resembles a ladder.

This diagram consists of one vertical line running down the left side, with the horizontal lines branching off to the right. The line on the left is called the bus bar, while horizontal lines are instruction lines or rungs. Along the instruction lines conditions are placed, that lead to other instructions on the right side. Power flow is always from left to right.

Therefore, the logical combination of these conditions from left to right side determines when and how the instructions at the right side are executed.

In a ladder logic diagram, instruction lines can have multiple branches. The vertical pairs of lines are called conditions. Conditions without diagonal lines through them are called NO conditions that correspond to AND, LOAD, or OR instruction.

The conditions with diagonal lines through them are called NC conditions that correspond to AND NOT, LOAD NOT, or OR NOT instruction.

Each condition has a number above/below each condition that indicates the operand bit for the instruction. Operand bit (Input/ Temporary bit) is associated with that condition.

The status of the bit determines the execution condition for the following instructions.

2.4.3 Basic terms used in ladder logic

Normally open and normally closed conditions

Each condition in a ladder logic diagram is either 'ON' or 'OFF' depending on the status of the operand bit that has been assigned to it. A NO condition is 'ON' if the operand bit is 'ON' and it is 'OFF' when the operand bit is 'OFF'. On the other hand, a NC condition is 'ON' if the operand bit is 'OFF' and it is 'OFF' when the operand bit is 'ON'.

In short, an NO condition simply follows the bit status (on => on and off => off) and an NC condition follows inverted bit status (on => off and off => on).

Execution conditions

In a ladder logic program, the logical combination of 'ON' and 'OFF' conditions before an instruction determines the conditions under which the instruction is executed. This condition is called the execution condition for the instruction. Except for the 'LOAD' instruction, all other instructions have execution conditions.

Operands

The operands designated for any of the ladder logic instructions can be I/O bits, flags, work bits or flags, timers, or counters, etc. In a ladder logic diagram, these conditions can be determined using these operands.

Logic blocks

The manner in which the conditions correspond to instructions is determined by the relationship between the conditions, within the instruction lines that connect them. Any group of conditions that go together to create a logic result is called a logic block.

2.4.4   Ladder logic instructions

The ladder logic instructions correspond to the conditions on the ladder logic diagram. Ladder instructions are either independent, or are in combination with the logic block instructions, from the execution conditions, based on which the execution of all other instructions are dependent. The most common ladder logic program instructions and the symbols used are shown in the Figure 2.11.

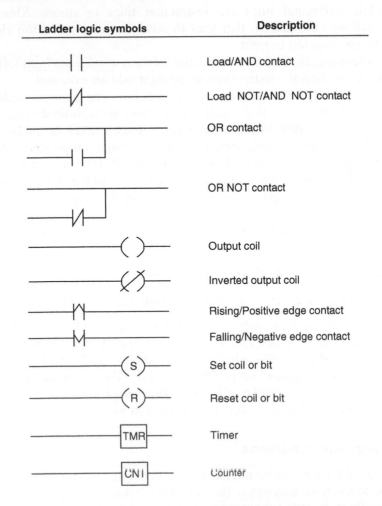

Ladder logic symbols	Description
	Load/AND contact
	Load NOT/AND NOT contact
	OR contact
	OR NOT contact
	Output coil
	Inverted output coil
	Rising/Positive edge contact
	Falling/Negative edge contact
	Set coil or bit
	Reset coil or bit
	Timer
	Counter

Figure 2.11
Commonly used ladder logic program instructions and symbols

2.4.5 The 'END' instruction

The last instruction required to complete a ladder logic program is the 'END' instruction. When the PLC CPU cycle runs through the program, it executes all the instructions up to the first 'END' instruction. After the 'END' instruction, it returns to the beginning of the program and begins the execution again. Usually, the 'END' instruction is the last instruction in the ladder logic program, but it can be placed at any point in the program, like when program debugging is undertaken. No instruction after the 'END' instruction is executed. The 'END' instruction requires no operands, and no conditions can be placed with the 'END' instruction.

2.4.6 Examples of simple ladder logic instructions

Examples of ladder logic instructions for simple control circuits (AND, OR, AND with OR) are shown in Figure 2.12.

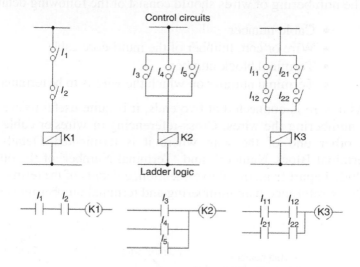

Figure 2.12
Examples of ladder logic instructions for simple control circuits

2.4.7 The ladder logic diagram

The ladder logic diagram is one of the methods of programing PLCs and is covered in detail in the IEC standard 61131 Part 3. Ladder diagram is a very convenient way of representing interlocking logics which used to be configured using hard wired devices. Current generation PLCs have other capabilities including that of PID controllers. The IEC standard therefore provides more advanced methods of programing such as Structured Text, Function Block diagram and Sequential Function chart for tasks which cannot be adequately represented using Ladder diagrams alone.

2.5 Wires and terminal numbering

In any electrical control panel, there are wires to which various electrical devices are connected.

It is important that electrical devices in a circuit are connected accurately through wires with proper voltages and polarity.

To ensure proper connections of wires, devices as well as terminals (through which they are routed) are given unique numbers.

This practice is followed for designing, assembling, and maintenance. This helps in identifying the devices, wires, and terminals during troubleshooting.

In an electrical panel, terminals are used to connect the wires. Generally, they are grouped together and called 'Terminal Block'. They are grouped either as per their functional use or as per the device connected.

Each terminal block consists of a group of terminals with an assigned 'Terminal Block Number'. Each terminal on the block is assigned a unique 'Terminal Number'.

In a panel, usually one side of the terminal is used for connecting internal wires from the devices inside the panel, and the other side is used for field or external connections.

In electrical panels, wires and cores of multi-core cables are used for interconnections. Wires and cable cores are terminated on device terminals and terminal blocks. Wires and cable cores used for interconnection are numbered. Alphabetical symbols and numbered ferrules are used on each wire or core of the cables.

The numbering of wires should consist of the following details:

- Cable number
- Wire or core number of the multi-core cable
- Terminal block number
- Terminal number on which the wire is to be terminated.

As a wire is connected at two ends, it is quite useful to use a cross-referencing method for numbering the wires. Cross-referencing of wires or cable cores include the details of the other end of the wire where it is terminated. Details such as 'Panel Number', 'Terminal Block Number', and 'Terminal Number' of the other end of the wire are also included apart from the above-mentioned details of the termination end.

Cross-reference wire numbering and terminal numbering is shown in Figure 2.13.

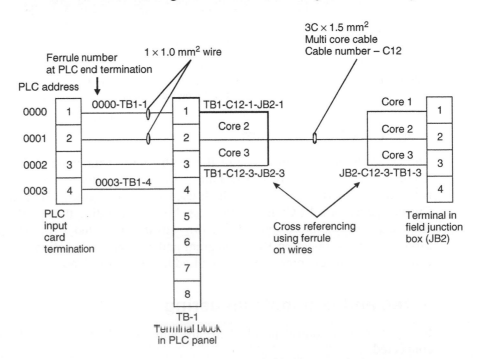

Figure 2.13
Cross-reference numbering of wires and terminal numbering

Though the wire numbering and terminal numbering shown in the figure is typical, in practice, there are many ways and methods of numbering wires and terminals that may be adopted. A cross-referencing number is one of the methods found useful during cable laying and termination, continuity testing, and troubleshooting.

As shown in Figure 2.13, cross ferruling is used between TB1 and JB2 terminal blocks for cable C12 cores. The ferrule at TB2 terminal block gives an idea of where the other end of the core is connected.

For example, as shown in the Figure 2.14, a PLC panel wiring along with cross-reference ferruling, PLC address information is also included. It is quite useful to include the PLC address in the wire ferrule number apart from the cable number, core number, terminal number, and cross-reference detail for troubleshooting.

In Figure 2.14, cross-referencing ferruling is used for field devices and terminal block wiring, as well as inter-terminal block wiring. Although this kind of ferruling involves lengthy ferrule numbers, the practice is certainly worth the effort while troubleshooting.

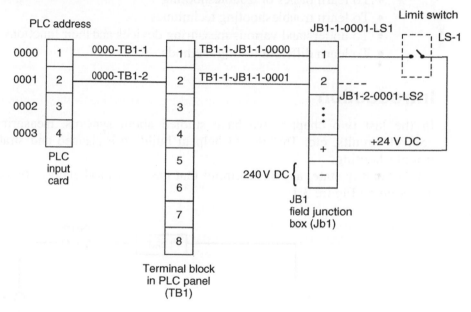

Figure 2.14
Wire numbering in a PLC panel with additional details such as PLC addresses

3

Basic troubleshooting principles

Objectives

- To learn basics of troubleshooting
- To learn troubleshooting techniques
- To understand various measuring devices and their functions
- To learn different testing methods.

3.1   Introduction

In the last two chapters we have studied about symbols, measuring meters, reading control circuits, etc. This would help in building a planned and strategic approach for troubleshooting.

In a healthy state, any basic circuit that has a load and energy source has circuit paths, as shown in Figure 3.1.

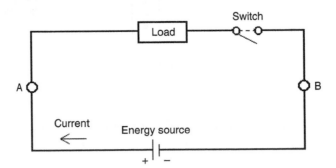

Figure 3.1
A healthy electrical circuit.

Current flows in a closed circuit between two electrically unequal potential points. Points A and B are any two points across which voltage is measured. The conductor offers resistance to this flow of electrons (i.e., current) depending upon the material.

Generally electrical problems can be classified under two broad types.

1. A connection does not exist where it should. This is an open circuit fault and can be detected using a continuity tester (Figure 3.2 illustrates this type of fault).

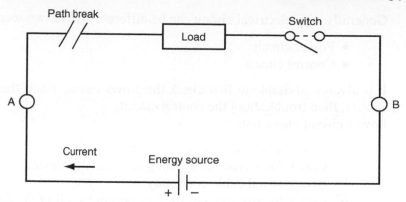

Figure 3.2
A typical open circuit fault

2. A connection exists where none should. This is called a short-circuit fault and can lead to excessive current accompanied by mechanical forces and heating of circuit conductors. Such fault happen due to insulation failures and can be detected using insulation testing instruments.

The process of detecting these faults and rectifying the circuit to restore normal operating condition is called troubleshooting. We will discuss about the open circuit fault first.

Current has a tendency to flow between two points that are at an unequal potential (electrically), provided the path between the two points is electrically conductive.

- Resistance offered by the path is known as 'Resistance' ohm)
- Electrical potential is denoted as 'volts'
- Flow of electrons between two points is termed as 'Current' (amp).

Therefore, while troubleshooting, the following points have to be checked:

☑ Continuity of path (i.e., resistance)
☑ Electric potential at two points of the path (i.e., voltage)
☑ Flow of electrons through the path (i.e., current).

An electrical circuit is made up of different paths and works on different voltages. Let us, therefore, identify the path that should be complete, also, when and how it is completed. Moreover, if it does not complete, let us identify the reason.

3.2 Basic principles in using a drawing and meter in troubleshooting circuits

To identify a faulty section, follow the guidelines given below, along with a drawing and a meter:

- Check the incoming supply voltages first
- Check for voltages at the specific test points in circuit (as per manufacturers test point data sheet)
- Do dead test of circuit for integrity of protection devices and others
- In dead test, check for continuity of circuits, as intended, and check for insulation resistance
- If it's not possible to perform a dead test, connect the supply to the circuit and do a live test of circuit.

Generally, any electrical circuit can be differentiated in two sections:

- Power circuit
- Control circuit.

It is always advisable to first check the power circuit. So, if the power circuit works, as it should, then troubleshoot the control circuit.

Power circuit check list:

- ☑ Incoming power to circuit and its integrity
- ☑ Check for correct functioning of protection devices
- ☑ Check visual cable continuity
- ☑ Check for any signs of flash or burning smell of devices.

Control circuit check list:

- ☑ Control circuit power first
- ☑ Check for proper functioning of relays, timers, and switches
- ☑ Check visual cable continuity
- ☑ Check for wire interconnections and terminal connections of circuit
- ☑ Check logical operational sequence of contactor switching
- ☑ Check for timer duration settings.

If the above criteria are checked and still the motor (final device) is not working, then test the motor (final device).

3.3   Checks for circuit continuity with disconnected supply

Dead circuit testing is testing performed with the power disconnected from the circuit. The main benefit of disconnecting power supply while tests with an external energy source are performed is to eliminate hazardous risks to the environment or the person conducting the test. A continuity test, as well as, an insulation test can be performed in the dead circuit test.

(a) *Continuity test*

This is to be performed on a dead circuit for checking continuity. Using an 'Audible Continuity Tester' can do it. This tester consists of a battery as a source of energy, an audible device, and two test leads. Figure 3.3 shows an example of this test with an audible continuity tester.

By this test, the continuity of an electrical circuit is checked to ensure that the electrical path is complete.

If the path is continuous, then an audio sound is emitted to confirm path continuity and the non-existence of an open circuit.

In some devices, along with the audio indication, an LED or some other visual indication is provided.

Similarly, an ohmmeter or multimeter can also be used to check continuity. An ohmmeter or multimeter consists of a battery as a source of energy, along with a meter to display the value of resistance. Figure 3.4 shows an example of this test with an ohmmeter.

In an ohmmeter, the scale is calibrated from zero to an infinite range of resistance. When the meter shows a zero reading, it indicates that the path between two test leads has zero resistance. This, in turn, indicates that the path is a continuous one. If the path or the conductor is open, then it will show resistance value as infinite.

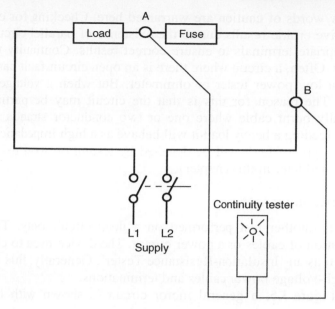

Figure 3.3
Continuity test with audio tester

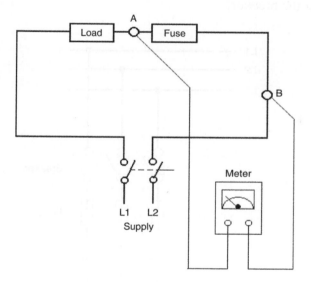

Figure 3.4
Continuity test with ohmmeter

In short, continuity testing is used to check the following purposes:

- Integrity of cables
  - ✓ Integrity of electrical circuit path
  - ✓ Integrity of the earthing system (i.e., electrical continuity and low-resistance value to earth)
  - ✓ Accurate wiring of a control and power circuit to the correct terminals
  - ✓ Differentiate active and neutral conductors before connecting them to a device
  - ✓ Check for wrong wiring interconnections between different control and power circuits; thus indirectly, checking for short-circuit paths
  - ✓ Integrity of switches, fuses, and other devices.

A few words of caution are warranted here. Checking for continuity in a control circuit can give erratic results due to the existence of parallel circuits. It is better to disconnect appropriate terminals to ensure correct results. Continuity test in power circuits can be tricky. Often, a circuit where there is an open circuit fault can register excellent continuity with a low power tester or ohmmeter. But when a voltage is applied, current may not flow. The reason for this is that the circuit may be partially continuous (Example: a partially burnt cable where one or two conductor strands may be making contact) but when feeding a heavy load it will behave as a high impedance.

This type of fault will be detected by testing on load using voltage measurements (as illustrated later in this chapter).

(b) *Insulation test*

This is another test performed on a dead circuit only. The objective is to check for insulation of cables or a power circuit. The device used to check integrity of insulation is known as an 'Insulation-Resistance Tester'. Generally, this is used during the installation of high-voltage power cables and terminations.

In Figure 3.5, a general motor circuit is shown with breaker, fuses, and overload relay.

To check insulation of the circuit (excluding motor), disconnect the power supply by opening the breaker.

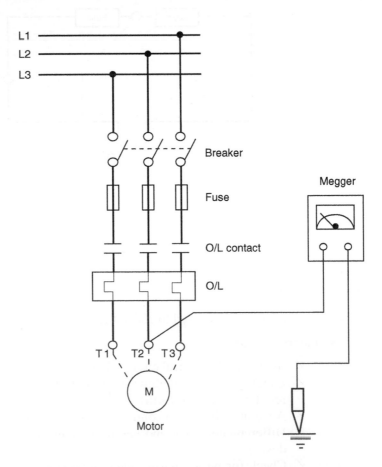

Figure 3.5
Insulation test with insulation-resistance tester

Then, isolate the motor from the circuit through terminals T1, T2, and T3. First checking insulation resistance between earth and T1, then earth and T2, and finally earth and T3 checks insulation resistances of conductors, as well as other devices.

If the insulation resistance of any branch shows zero or a very low reading, then it can be concluded that there is an insulation failure.

This test is also used in fault finding, to check for earthed motors or cables and for checking insulation failure of conductors. Individual phases of three-phase motor winding can be insulation-tested only if all six leads of the winding are brought out. The winding being tested should be connected to the tester's output with the other two windings connected together and to the earthed frame of the motor. Where only three leads are available, the insulation of the machine winding as a whole can only be tested with reference to the earthed frame of the motor.

These insulation testers are also called Meggers and have a built-in energy source (either DC generator or battery) to produce test voltages of rating 500 V DC or more. This is required since the electrical circuit to be tested applies voltage of different ratings.

For example, when the insulation resistance of HV cables is checked, 1000 V minimal voltage is applied, whereas for a domestic circuit 500 V is sufficient for testing.

Testing on a live circuit requires extreme caution and should be restricted to LV circuits.

Precautions should be taken to prevent inadvertent contact of the technician with live parts.

The probes and tools must be insulated with minimum exposure of conducting parts. This will minize inadvertent bridging of two terminals which are at different potentials which can cause a short-circuit and arcing leading to burn injuries to the technician.

3.4   Checks for circuit continuity with live supply

Generally, if possible, troubleshooting is done with a disconnected power supply, but in some circumstances, faultfinding is only possible if the circuit is live.

Therefore, the circuit under testing remains connected with the power supply. This uses the circuit power supply itself as a source of energy for testing.

This kind of testing should be done with extreme care following safety precautions.

As shown in Figure 3.6, the integrity of a power supply or continuity of electric path can be checked by using test lamps. Test lamps are connected in between two phases. Thus, as with the dead circuit test, a continuity test can be performed. In addition, a lamp-type visual tester can be used for simple continuity testing. Alternatively, voltage indicators or multimeters can be used for checking voltage and the continuity of the conductors or electrical path.

While checking three-phase voltage, use two lamps connected in series and not a single lamp. Currently, most manufacturers give test voltage details for test points that helps to check the integrity of a particular section. Generally, equipments consisting of electronic cards follow this kind of practice.

While checking the voltage at these test points, measuring instruments must be accurate. Therefore, a comparison of voltages at these test points is sufficient to draw conclusions.

Diagnostics for a single-phase motor can be undertaken with this kind of visual indicators. This requires a sound knowledge of circuit and wiring arrangements – depending upon the test done, interpretation varies and so does an accurate fault diagnosis.

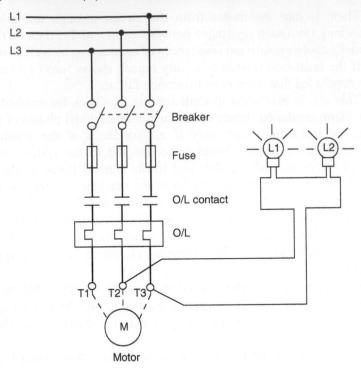

Figure 3.6
Continuity test with series test lamps

It is always advisable to check voltage between line-to-line than line-to-earth, since for the latter, the results may be misleading.

Testing on a live circuit requires extreme caution and should be restricted to LV circuits.

Precautions should be taken to prevent inadvertent contact of the technician with live parts.

The probes and tools must be insulated with minimum exposure of conducting parts. This will minimize inadvertent bridging of two terminals which are at different potentials which can cause a short-circuit and arcing leading to burn injuries to the technician.

3.5   Tests and methods

Although most of the tests have been detailed above, there still remain certain tests that are common and critical to an electrical system. Parameters such as the resistance of leakage paths and resistance of a conductor have to be known for some applications.

The tests include the following:

☑ Effectiveness of the power earth
☑ Effectiveness of the electronic earth
☑ Continuity of the earthing system and the required equipotentiality (so as to check the possibility of earth current loops)
☑ Earth pit locations, resistance of earth conductor, material, and size
☑ Location of protection devices, so that the path taken by fault current is minimum and insures the activation of a protection device under fault conditions
☑ Ratings of fuse and other protection devices

☑ Selection of suitable cable types with proper current ratings keeping in mind environmental conditions and length of run

☑ Materials prone to environmental hazards which may be mechanical or chemical, which might be present, such as dampness, high temperature, explosive gases, vapors

☑ Electrical equipments, to ensure that its operation will not cause overload conditions

☑ Location of installation of electrical equipment and accessories.

If all the above points are considered during the installation of an electrical system, there will be a reduced need for troubleshooting. The quality of the system will also be good.

3.6   Testing devices

To successfully troubleshoot in a short time, an understanding of the measurement meters that can be used and their various functions is mandatory. This is detailed in the following few sections.

3.6.1   Lamp indicators

A lamp indicator is the most basic tool used for troubleshooting by a practicing electrician. It is also known as a 'Voltage Tester'. It consists of two 240 V lamps connected in series.

Description

As shown in Figure 3.7, both lamps are connected in series along with the fuse and probe that form a testing set. The low-wattage lamps comprise of equal power rating, not greater than 25 W per lamp. Use of two lamps is advisable as the tester may at times be subjected to line voltage (380/400/480) during testing. Single lamp if used may fail and erroneously indicate that the circuit is not live. As a precaution, all voltage testers should be checked before and after the test using a known live source.

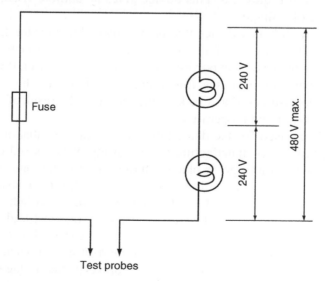

Figure 3.7
The two 240 V lamps housed in the housing

For testing extra low voltages such as 12 or 32 V systems, one may use a single-lamp type.

Applications

Lamp indicator applications are listed below:

- For detecting the presence of a live potential
- For polarity of supply, i.e., the location of active points, neutral, and earth terminals or supply points
- For checking like or similar phases when 'phasing out' preparatory to paralleling two supplies
- Blown fuses
- Integrity of motor and three-phase supply system.

Testing of a motor

To check the earth condition, one lead of test lamps is connected to a live terminal of a single-phase supply and the second lead to a winding terminal. If the winding is earthed, the lamps will glow, else they will not glow.

Checking a three-phase supply voltage

To ensure there is no missing phase, connect both leads across the two phases. One of the following three instances will occur:

- If there is no supply then lamps will not glow
- If any one phase is missing then the lamps will glow at half brightness
- If both phases are connected then the lamps will glow at full brightness.

Present day voltage testers (duly approved types), when correctly handled, are safe live-testing devices due to their impedance and rating that a user cannot even mistakenly cause a short-circuit.

One such voltage tester available in the market is known as the 'high-impedance tester', as shown in Figure 3.8. This device gives an audible visual indication on detection of the presence of voltage.

A common handy device that is used by technicians for detection of voltage is the 'Neon Tester'. This consists of a neon indicator with current-limiting resistor in series. When you put its conductive front portion over an active conductor, it will give an indication.

However, in most cases, this indication is faint and this confuses the user. In addition, it cannot be assumed that a lack of indication is the result of a lack of supply. In that case, a neon bulb may be inoperative.

A device used for the detection of electrical potential or for a polarity test is the 'neon test pencil'. It is manufactured in a variety of types and designs. The intensity of glow will increase if a finger is placed on the cap or if the cap is earthed.

Therefore, it is advisable to use a good-quality neon tester from a safety and reliability point of view. It should be noted that the neon tester indicates the voltage of a conductor with reference to earth (because it has only one test lead and the circuit gets completed through the body of the person using it and earth). This test is sometimes not conclusive because an open circuit may exist in the neutral and remain undetected by this test. Also, neon testers sometimes give a glow at very low voltages and results are thus erratic. A proper voltage measurement using a voltmeter or multimeter is always more reliable and conclusive.

Figure 3.8
High-impedance tester

A comparison between analog and digital instruments is as follows:

- An analog meter shows reading by the movement of a pointer on a calibrated scale, whereas a digital meter shows a digital readout for the measurement directly.
- It is easier for a user to differentiate between readings in analog meters rather than in the digital one.
- In an analog meter, the reading is subjected to parallax error, while in a digital one there is no such possibility.
- For testing continuity, analog meters are a better choice.

For example, for an open-circuit full-scale deflection of the needle of an ohmmeter will show infinite reading. While for a closed circuit, it will show zero-resistance reading, thereby misleading the user.

3.6.2   Voltmeters and ammeters

For the measurement of voltage (potential difference) between two points, a device known as a voltmeter is used. This is a device used in live testing of a circuit.

Voltage measurement is conducted by connecting the voltmeter across the test points in a circuit.

A voltmeter can be used to measure AC/DC voltages of different ranges. Therefore, AC voltages should be measured by selecting AC and vice versa.

A voltmeter is always connected in parallel or shunt with respect to test points.

While operating a voltmeter, ensure that proper range of voltage is selected before conducting the measurement, because an instrument is designed for a particular range.

Failure to maintain the above-mentioned precautions results in safety hazards to both the user and the instrument.

When high-voltage measurements are required, then the measuring range of a voltmeter can be extended further by the addition of a voltage transformer (step down) with the meter.

Accordingly, the measurement scale requires a multiplication factor.

Currently, digital multimeters (DMM) with voltage measurement have auto range facility. This enables the instrument to automatically get the correct range in spite of the user's incorrect selection of range.

A voltmeter is used for the following purposes:

- Test continuity of power in a electrical circuit
- Check integrity of single-three-phase power
- Check integrity of devices such as relays and timers
- Check integrity of earthing.

An ammeter is the other device used to measure current flowing through closed low-voltage electrical circuits.

In addition, it is used in the live testing of an electrical circuit. Connecting the ammeter in series with the close electric circuit always does measurement of current. AC/DC currents of different (LV) ranges can be measured using an ammeter.

When connected in series with load (motor, fan), an ammeter will indicate the current consumed by the load. The current shown depends upon the exact connection of the ammeter.

In order to extend the range of measurement done by an ammeter, a CT is connected along with the meter. Accordingly, a multiplying factor will come into the picture.

Figure 3.9 shows the connection of an ammeter and voltmeter used for troubleshooting a motor starting circuit. As explained earlier, an ammeter is connected in series with the path, while a voltmeter is connected across the test points.

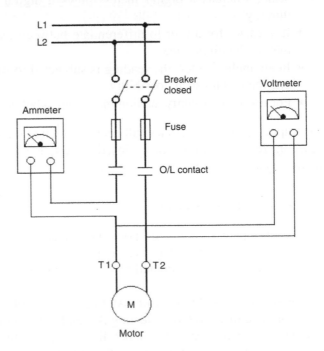

Figure 3.9
Simple circuit showing metering of current and voltage

For very-high voltage circuits, it is unfeasible to disturb the circuit or take physical connection risks with the meters.

To avoid physical connection of an ammeter with the circuit, another handy device known as the clip-on meter is available in the market (Figure 3.10).

Figure 3.10
Digital-type clip-on meter

A clip-on meter, as the name suggests, is a handheld device that requires to be clamped with an active current carrying conductor in a circuit (Figure 3.11). The basic principle remains the same – a CT transfers high-rating current to a low-rating meter, which shows the reading on a calibrated scale.

Figure 3.11
Using a clip-on meter

This device works for AC/DC ranges with an option of various current ranges. It also has a hold facility, which helps in value storage after a reading is taken. Moreover, this device has the facility to be used as a voltmeter. It can be turned into a voltmeter by using the extra probes provided for testing. Therefore, it is a versatile measuring device. It can be used for rough and ready measurement of current flow in locations where individual phase leads are accessible. However, it cannot measure current in multi-core power cables where all three-phase conductors run bundled into a single cable. When using a clip-on ammeter care should be taken to see that the CT clamp is fully closed without any air gap because the readings can be quire erroneous in the event of improper closing of the clamp.

3.6.3   Multimeters and ohmmeters

To perform various tests to check AC/DC voltage, current, resistance, frequency, continuity of circuit, or device integrity, the multimeter is a very useful device (Figure 3.12).

Figure 3.12
Analog multimeter

Various companies have different models with different functions. A multimeter consists of an ammeter, voltmeter, and an ohmmeter combined, with a function switch to connect the appropriate function.

The ohmmeter is essentially a current-measuring device. However, the scale is calibrated in ohms, enabling resistance values to be read directly.

This combination of volt–ohm–milliammeter is a basic tool for troubleshooting. The proper use of this instrument increases its accuracy and life. The following precautions should be observed during its usage:

1. To prevent meter overloading and possible damage when checking voltage or current, start with the highest range of the instrument and move down the range successively.
2. For higher accuracy, the range selected should be such that the deflection falls in the upper half of the meter scale.
3. Verify the circuit polarity before making a test, particularly when measuring DC current or voltages.
4. When checking resistance in circuits, power supply to the circuit has to be switched off; otherwise, the voltage across the meter may damage the meter.
5. Renew multimeter batteries frequently to ensure accuracy of the resistance scale.

6. Recalibrate the instrument at frequent intervals.
7. Protect the instrument from dust, moisture, fumes, and heat.

Digital multimeter

The features of a multimeter (Figure 3.13) are listed as follows:

- Functionally easy
- Directly indicates the numeric value of measurement on an LCD display
- Measurement-variable – voltage, current, resistance can be selected using the function button
- Auto range feature provides automatic adjustment of internal circuits to appropriate current, voltage, or resistance
- Hold feature allows storage of reading of quantity measurement in memory for future viewing
- Auto polarity feature automatically displays + or – sign on display to indicate polarity of DC measurements
- Some meters also provide Min/Max value indication for measurement
- Peak hold feature holds the peak value of measured quantity
- Quick check features such as diode test, transistor test, capacitor test, etc. are also available.

Figure 3.13
Digital multimeter

Operating DMM

To operate the DMM, please perform the following steps:

- Before connecting the test probe that leads to the circuit, ensure proper function has been selected as per measured quantity.
- Check the correct insertion of test probes in proper plugs – this will avoid possible damage to the multimeter due to an incorrect function selection or an incorrect probe insertion.
- If the multimeter does not have an auto range feature then check for the range selector switch position – now, the variable can be measured.

If the measuring position/location is awkward, data can be stored using the hold function for later viewing. The data can be viewed even after the probes are disconnected from the circuit.

3.6.4   CRO (cathode ray oscilloscope)

The CRO measuring instrument may sound very familiar, as it is a very useful device. It is used for measurement of voltages (AC/DC) and display of waveforms by providing information on time duration, frequency, and their shapes.

In the following section, we will discuss the features and mode of operation of the CRO.

Features of CRO

Below are listed the various features of the CRO:

- It allows voltage (AC/DC) amplitude measurement and time period measurement from the waveform displayed on screen.
- Dual trace CRO allows the user to see two traces at a time on two different channels for comparison.
- Two sets of controls provide the facility to show time period differences, amplitude differences, and shape/distortion comparison.
- Storage oscilloscopes allow storage of waveforms for later analysis.
- Storage facility is very useful since it provides a cursor function, which shows the value of a measured variable at a particular instance.

Operating CRO

To operate the CRO, perform the following procedure:

- The power on switch is provided for on/off control.
- The measurement probe provided consists of two leads – one connected to the signal and the other, ground probe, connected to the ground of circuit.
- Turn on the CRO.
- Check the integrity of leads and the CRO by connecting the I/P probe to a test socket of 5 V square wave signal.
- While checking non-isolated signals (that are earthed) do not connect ground/earth to the CRO, else it may create a short-circuit at the input signal.
- Adjust both channels' vertical axis by placing AC\DC\GND signal in GND position.

- Place the function switch in the suitable signal function as required (AC\DC).
- Check the test probe selection, i.e., divide by 1 or 10 that allows signal attenuation.
- The intensity knob is used to vary the brightness of the trace.
- The focus knob is used to change the sharpness of the trace displayed.
- The Y shift allows you to shift the waveform displayed vertically (up/down).
- The X shift allows you to shift the waveform displayed horizontally left or right.
- The Volts/Div switch is used to vary the magnitude of the voltage variable displayed on the screen. It is calibrated in Volts/Div of the vertical scale. A control knob is provided in the center to adjust amplitude between calibrated settings.
- To find the amplitude of a signal multiply the Y-axis reading with the Volts/Div setting.
- The Time/Div is used to control the span of the X-axis.
- Physical markings between two points can be used to calculate the time span. The same time span can be used to measure the frequency of the waveform displayed.
- A control knob is provided at the center for the same purpose as in Volts/Div.
- To find the time duration of a waveform, measure the signal span reading difference. When this is multiplied by Times/Div it will give the time duration of the signal.

3.6.5   Safety standards for measuring instruments

While a person is conducting tests with an instrument on live line supply, there is a possibility of a sharp rise in voltage for a short duration. This may result in an arc or flash between measurement terminals of the testing device. In addition, if a heavy flash occurs, it may critically injure a person handling the instrument.

To safeguard the person using the measuring instrument and to classify the various instruments as per the application they are used in, the IEC has classified instruments in the following categories:

- *Category IV*:   Distribution systems, service connections, and primary over-current protection for larger installations
- *Category III*:   Three-phase and single-phase distribution within a premises
- *Category II*:   Appliances, lighting points, socket-outlets
- *Category I*:   Transient-protected electronic equipment.

The IEC Standard 61010 provides guidelines for manufacturers to follow safety norms for testing devices. It is to be noted that irrespective of their maximum voltage rating, a Category IV device provides a greater degree of transient protection than Category III, etc. The Category III device is suitable for most of the testing undertaken by electricians.

3.6.6   Insulation-resistance testers or meggers

Another common method of measuring resistances ranging 0–1000 MΩ is by using meggers or insulation-resistance testers. This is the usual ohmmeter with a battery used for voltage source.

This instrument is used to measure very high resistances, such as those found in cable insulations, between motor windings, in transformer windings, etc.

Normal multimeters do not provide accurate indications above 10 MΩ because of the low voltage used in the ohmmeter circuit. Meggers can apply a high voltage to a circuit under test and this voltage causes a current if any electrical leakage exists. This makes it useful as an insulation tester.

Some laboratory test meters have a built-in high-voltage source. The high voltage permits accurate high-resistance measurement, but such meters are usually not portable.

The megger is essentially a portable ohmmeter with a built-in high-voltage source. The built-in high-voltage source may be derived from a magnet-type DC generator or battery.

In a DC generator-type megger, a hand crank is used to turn the armature to produce voltages up to 500, 1000, and 2500 (depending on the model used).

An electronic battery-operated type of instrument is popular because it is light, compact, and can be held and operated in one hand, i.e., there is no generator to turn. High-testing voltage is produced by an electronic circuit, which uses an internal battery as an energy source. The instrument in Figure 3.13 has a range of 0–100 MΩ and infinity at a testing voltage of 500 V. It has the same voltage-measuring facility as the hand-driven insulation tester.

Resistance is directly displayed on the front panel digital display. A range of voltages can be selected while testing.

3.6.7   Testing accessories

Depending upon specific applications, special testing devices are used specifically to improve the accuracy and efficiency of testing an installation. Some of these devices are shown in the figures given here. Figure 3.14 shows a hand operated insulation tester (commonly called as Meg. Ohm Meter or 'Megger').

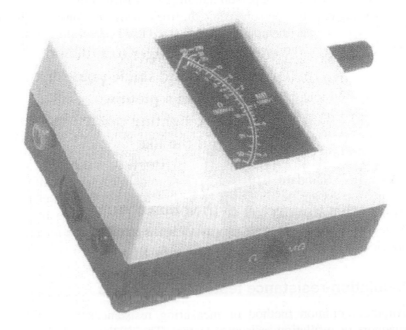

Figure 3.14
Hand-driven insulation and continuity tester

Figure 3.15 shows an electronic version of such a device, a multi-voltage digital insulation tester with the test voltage being generated from a battery source. Figure 3.16 shows an analog bersion of a similar instrument.

Figure 3.15
Multi-voltage digital insulation–resistance tester

Figure 3.16
Battery-operated insulation–resistance tester

Figure 3.17 shows various accessories forming part of the testing instruments. These instruments help electrical installation contractors to test their installation in order to ensure that the installation is safe to connect to the supply.

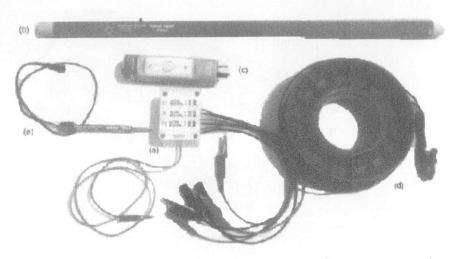

Figure 3.17
Installation tester

Figure 3.18 shows an instrument designed for the dead test of final circuits of an electrical installation.

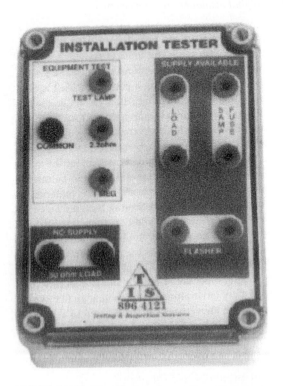

Figure 3.18
Installation tester

These devices are intended to help improve the safety and efficiency with which tests are carried out. The testing principles remain the same and these must be understood before the benefits of using such devices can be fully realized.

3.7 Circuits

The previous chapter has listed the symbols representing electrical and electronic components. Symbols make a drawing or circuit more readable.

Generally, a schematic consists of a control and a power circuit. The power circuit provides power to the motors through contactors, whereas a control circuit controls these contactors through safety interlocks.

3.7.1 Reading a circuit

- All control devices such as switches and relay contacts are either NO or NC contacts.
- The switch position normally shown in any circuit diagram is the default de-energized condition state.
- For denoting sensor contacts notations such as LS, PS, TS are used.
- A relay coil is denoted with a symbol inside a circle and contacts of relay used in the circuit are represented with the same tag as the coil. If a relay coil has multiple NO and NC contacts then the contact identification numbers, as shown on relays, are mentioned in the drawing.
- In between the control supply lines L1 and L2 you will find either the relay coil or a solenoid coil or the lamp load.
- If several devices are to be turned 'On' at the same condition then you will find them connected in parallel between L1 and L2.
- If wires are common for two devices then in the diagram they are shown with the same identification number.
- Generally, power circuit conductors are shown with thick lines while thin lines are used for control circuits.
- A broken line indicates a mechanical function. Generally, it is used to show linkage between two different contacts of the same push button.

3.7.2 Different wiring diagrams

In order to troubleshoot electrical equipments two things are required. One is the location of the equipment to be tested and the other is the interconnection between all the devices (contactors, timers, relay).

The wiring diagram of electrical equipment gives information as stated above. In addition, it shows the identification tags of wires, connectors, relays, etc.

In Figure 3.19, the wiring diagram of a DOL starter is shown along with the physical location of the devices. The terminal numbers of overload relays are also shown in the wiring diagram. This enables accurate device wiring and wire tracing during troubleshooting.

Generally, this kind of wiring diagram is given inside an electrical equipment panel cover.

This diagram shows the actual position of different devices as closely as possible. The bold line indicates the heavy current carrying conductors, while the thin line indicates the control circuit.

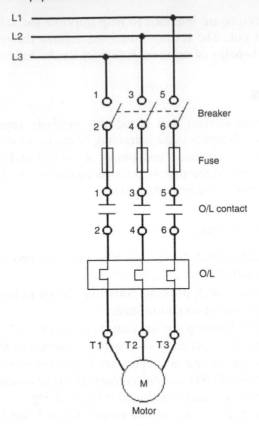

Figure 3.19
Simple motor circuit

3.8 Accurate wiring of circuits and connections

While troubleshooting electrical equipments, continuity tests of circuits and wiring are done by performing the following procedure:

- Checking correct polarity and ensuring that supply polarity follows the correct circuit route.
- Ensuring that there are no short-circuits in supply due to a wrong connection or termination of wires.
- Identifying different conductors before making connections to a device to ensure correctness of circuits and connections.
- Ensuring that there are no interconnections between two different circuits.
- Correctly identifying circuit loads such as contactors, relays, and their contacts.
- Identifying active conductors and their corresponding neutral conductors to check the integrity of the circuit.

A continuity test is particularly useful to help detect a short-circuit condition, which is a result of cross-linking of wires between two different circuits.

An interconnection between circuits is likely to be due to the following reasons:

- Incorrect termination of wires
- Result of insulation breakdown
- Incorrect connection at field junction box.

Figure 3.20 is an example of an electrical appliance connected with a supply system. If, due to any reason, a fault occurs within the appliance causing current flow in its body, the fault current flows back to the mains supply. The circuit shown here is a TN-C-S type of supply where the earth is derived from the supply neutral at the service entrance. In the case of other supply systems the earth lead may not be interconnected at service entrance but go right back to the source (TN-S). However the general principles are still valid.

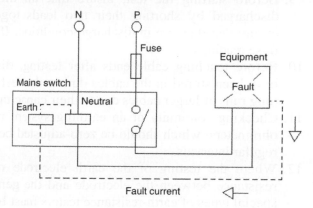

Figure 3.20
Earth fault within an installation

It is required to conduct tests between neutral conductors of all other circuits and the active conductor of the same circuit at the mains supply distribution to reveal any interconnection faults.

Before conduction of tests, perform the following steps:

- Disconnect neutral link from circuit
- Keep circuit protective
- Close all contactors or switches.

Check all direct interconnections with the low-range ohmmeter. If resistance shown in the ohmmeter is very low then it indicates a short-circuit condition. Suppose, the load is connected with an active phase and is neutral from different circuits, then it can be detected only with connected loads. If these steps are performed prior to the start of the test, then check the resistance between the neutral and the active conductors.

To check for insulation resistance of cables, take insulation resistance with megger or insulation-resistance tester, especially if insulation breakdown is suspected. If the resistance shown is less than 1 MΩ then it can be said that the wiring or device terminal has an insulation problem.

To identify each electrical circuit and its active and neutral conductors, calculate load resistance with the ohmmeter and accordingly, identify each active and neutral conductor.

Conducting an insulation-resistance test:

To conduct an insulation-resistance test, perform the procedure listed below:

1. Check the insulation tester by shorting its test leads. It should show zero resistance. If test leads are kept open, it should show infinite resistance.
2. Isolate the section to be tested from the power supply.
3. Disconnect all lamps or electronic devices from the circuit to be tested.
4. Select the proper operating voltage for conducting the test, depending upon the rating of the system.

5. Check for connections while conducting tests so that only the section to be tested is included in the test.
6. There should not be any stray parallel leakage paths.
7. Check the instrument for pointer index or any other pre-adjustment necessary.
8. Test leads to be used should have good-quality insulation.
9. Before starting the test, insure that all the capacitors in the circuit are discharged by shorting their two leads together. Similarly, after the test ensure that they are in discharge condition. If this is not done they may give false readings.
10. Before touching cable ends after testing, discharge any energy that might have been stored in the cables during the test. This is most likely to occur in long runs of larger cables due to their capacitance.
11. Checking continuity of an earthing system requires the use of low-reading ohmmeters, which should be zero-adjusted before each test and calibrated on regular intervals.
12. Where the testing of the earth electrode resistance is required (i.e., the resistance between the electrode and the general mass of earth), one of the special types of earth-resistance testers must be used.

3.8.1   Optional tests

There still remain a few useful tests using the measuring devices.

These tests are used for checking single- or three-phase systems and other electrical devices. Some of the tests we have discussed in the next few sections. These can be used to strengthen troubleshooting techniques.

(a) *Megger testing cables and auxiliary devices of a single-phase system*

Disconnect P and N from the supply side, as well as from the other end.

Now we have isolated our test circuit, making it dead. Short P and N with a temporary short link. Close switch and protection devices.

As shown in Figure 3.21, open motor terminals, so that the motor remains isolated from the test circuit. Check resistance with the insulation tester between the neutral link and earth. If the value shown in the meter is less than 1 MΩ, then there is a fault with either the cable insulation or device terminals.

(b) *Megger testing cables and auxiliaries of a three-phase system*

Disconnect L1, L2, and L3 from the supply side, as well as from the other end. This makes it a dead circuit.

Short L1, L2, and L3 terminals with a temporary link. Close the breaker device and protection devices. As shown in Figure 3.22, open motor terminals T1, T2, and T3, so that the motor remains isolated from the test circuit.

Check resistance with insulation tester between each conductor and earth.

If the meter shows a low value less than 1 MΩ, there is a fault in either cable insulation or device terminals.

(c) *Meggor testing of motor*

A pre-condition for megger testing of a motor is to isolate the motor from the supply totally. Take the megger value of a motor between each conductor and earth, as shown in Figure 3.23, to check the earthing of the stator winding. This will help us to conclude on earthing status of the stator winding. Similarly, check for shorting between two windings by checking the megger value between two stator-winding terminals, as shown in Figure 3.24.

Thus, a low reading can identify insulation failure of any winding inside the motor.

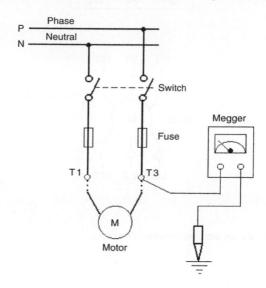

Figure 3.21
Megger of a single-phase system

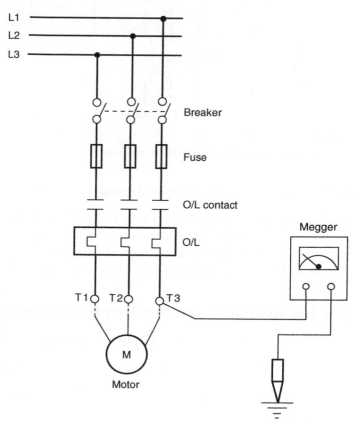

Figure 3.22
Megger of a three-phase system

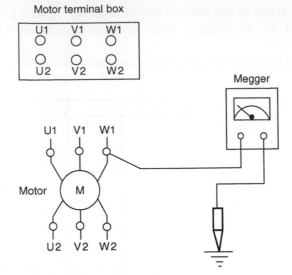

Figure 3.23
Megger testing for an earthed winding condition

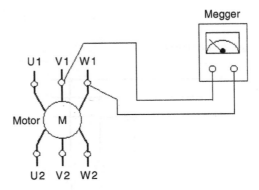

Figure 3.24
Megger testing for winding-to-winding short condition

Fault finding on an underground cable

Generally, underground cables are prone to insulation failure, although due care is taken. Since they are buried underground, it is difficult to exactly pinpoint the fault location. Resistance values between earth and a conductor from the two ends can be checked. If the value reduces drastically, then the faulty location can be isolated.

3.9   Tests for installation and troubleshooting

The following are a few tests used during commissioning and troubleshooting:

1. *Insulation test*:   This is the most important test for troubleshooting of any electrical equipment. Depending on the system, a suitable test voltage is applied to check the insulation resistance between the live conductors and earth.
2. *Earth continuity test*: For electrical equipment, continuity between the exposed portion (metallic) of earthed equipment and the earth terminal is

checked. Resistance value should be low. If resistance value is high, then it is indicative of poor earthing.

3. *Flash test*:   To check the insulation strength of cables, a high voltage (as specified by the cable manufacture) is applied in the same way as for the insulation test. This determines the withstand capability of the cable insulation.

4. *Electronic earth test*:   Generally, for microprocessor-based or electronic-based sensitive devices, separate earthing is provided. This is called electronic earth. The voltage between the electronic earth and the power earth should be lower than 2 V.

4

Troubleshooting AC motors and starters

Objectives

- To understand the basics of single-phase and three-phase AC motor
- To understand the basics of DC motors
- To explain connection details and braking methods of motors
- To explain testing methods of motors.

4.1   Introduction

AC motors provide the motive power to lift, shift, pump, drive, blow, drill, and perform a myriad of other tasks in industrial, domestic, and commercial applications. The induction motor, the most versatile of the AC motors, has truly emerged as the prime mover in industry, powering machine tools, pumps, fans, compressors, and a variety of industrial equipments.

This section begins with the fundamentals of AC motors, to provide a sound base for understanding the practical aspects of induction motor applications in industry. It aims to impart, through a graduated approach, the necessary cognitive and technical inputs to diagnose and troubleshoot AC motors and starting gear and develop a 'preventive' approach to optimize motor performance, reduce downtime, and extend operational life.

The following sections first detail the three-phase AC motors, then single-phase AC motors, and then DC motors.

4.2   Fundamentals of three-phase AC motors

Three-phase AC motors are known as the 'workhorses of industry' because of their wide use and acceptance. They are popular because they are low in cost, compact in size, require less maintenance, withstand harsh industrial environments, etc.

Three-phase AC motors are a class of motors that convert the three-phase electric power supplied at the input terminals, to mechanical power at the rotating shaft, through the action of a rotating magnetic field, produced by a distributed winding on the stator.

Three-phase AC motors are broadly classified as:

1. Induction motor
2. Synchronous motor
3. Wound rotor induction motor.

Each motor operation is detailed briefly.

1. *Induction motor*

As the name implies, no voltage is applied to the rotor. The voltage is applied to the stator winding and when the current flows in the stator winding, a current is induced in the rotor by transformer action. The resulting rotor magnetic field will interact with the stator magnetic field, causing torque to exert on the rotor.

2. *Synchronous motor*

As the name suggests, rotor speed remains in synchronism with that of the stator magnetic field. The motor runs at the same speed.

Unlike induction motors, synchronous motors are not self-starting. They have to be brought up to synchronous speed. Once they are locked then the rotor will continuously rotate.

3. *Wound rotor induction motor*

This motor has a 'wire wound rotor' from which three leads are brought out to the slip rings. It is possible to vary the rotor resistance. Introducing different resistances in the rotor circuit through the slip rings does this. The speed and the starting torque will now be variable.

4.2.1   Principles and operation of three-phase induction motors

Three-phase induction motors have three coils, placed 120 electrical degrees apart from each other, which form the stator winding.

Whereas a rotor is a squirrel-cage type (solid one) and has copper conductors, which are shorted at one end by a circular connecting plate. In Figure 4.1, a squirrel-cage induction motor is shown.

When voltage is applied to the stator winding, current flows through it, creating a rotating magnetic field. The speed of this rotating magnetic field depends on the number of poles of the stator, and the frequency of supply given to it. This is known as Synchronous speed and is given as:

$$S = \frac{120\,f}{p}$$

Where
  S = Synchronous speed in RPM
  f = Frequency of source in Hz
  p = Number of poles of stator winding.

The rotating magnetic field induces emf in the rotor by the transformer action. Since the rotor is a closed set of conductors, current flows in the rotor. The rotating fields due to stator currents react with the rotor currents, to produce forces on the rotor conductors and torques.

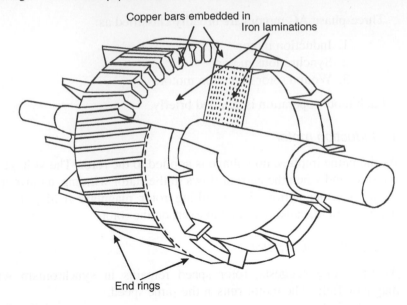

Figure 4.1
Squirrel-cage induction motor

This motor is called an induction motor, as it works on the principle of transformer action or induction.

Characteristics of a three-phase induction motor:

- No external starting mechanism is required.
- They come in a variety of horsepower ratings.
- Speed is inherently constant.
- The direction of rotation can be changed easily by reversing any two power lines of the motor.
- Motor runs at a speed lower than the synchronous speed by a factor 'slip'.
- For reduced loads, the power factor becomes poor.
- Multi-speed squirrel-cage motors are available with a provision of changing the number of stator poles by changing external connections.

4.2.2   Speed–torque characteristics of an induction motor

The speed–torque characteristics and torque–slip characteristics of an induction motor are important parameters for determining the performance of the motor. Typical speed–torque characteristic and torque–slip characteristics of a three-phase induction motor are shown in Figure 4.2.

It can be seen that when a motor starts from zero speed, the start torque is lower than the full load torque and the motor can start at light-to-no load.

The normal full load torque is achieved at a point where the rotor speed is only 5% less than the synchronous speed. From this point onwards, the torque drops to zero value since there is no relative motion or slip between the stator and the rotor.

In order to achieve a high starting torque, the rotor is made with high-resistance conductors or else an external resistance is inserted in the rotor circuit.

The nature of the characteristic curve can be changed, in case of a slip-ring type induction motor, by inserting an external resistance in the rotor circuit. If the rotor resistance is increased from r_1 to r_2, r_3, r_4 ($r_1 < r_2 < r_3 < r_4$), then the maximum torque remains the same, but the slip at which the maximum torque occurs is shifted, as shown in Figure 4.2. The method of introducing an external resistance in rotor circuit is used, to obtain a higher starting torque, as required, up to the maximum torque limit that the motor can produce. This method of increasing the starting torque can be used only in the case of slip ring or wound rotor induction motors.

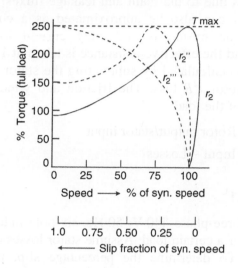

Figure 4.2
Speed–torque and torque–slip characteristics of an induction motor

4.2.3   Induction motor startup

The main objectives while starting an induction motor are:

- To handle high-starting current
- To achieve high-starting torque.

As discussed earlier, rotor resistance determines starting torque. Usually, this rotor resistance is small, giving small starting torque, but good running conditions. So, the squirrel-cage motor can run only with low-starting loads.

If the rotor resistance is increased by some means, then the slip and speed at which maximum torque occurs can be shifted. For that purpose, external resistance can be introduced in the rotor circuit, which is done in the case of slip ring or wound rotor type motors.

When power is applied to a stationary rotor, excessive current will start flowing. This happens due to the fact that there is a transformer action between the stator winding and the rotor winding, and the rotor conductors are short-circuited. This causes heavy current flow through the rotor.

If, for reducing this heavy starting current, starting voltage applied is reduced then it affects the starting torque as well.

To get everything out, the following method of starting is generally used:

- DOL starting
- Auto transformer starting
- Star–delta starting.

4.2.4   Induction motor losses and efficiency

The following are the losses in an induction motor:

- Core loss in the stator and the rotor
- Stator and rotor copper losses
- Friction and windage loss.

Core loss is due to the main and leakage fluxes. As the voltage is assumed constant, the core loss can also be approximated as a constant. DC can measure the stator resistance. The hysteresis and eddy current loss in the conductors increase the resistance, and the effective resistance is taken at 1.2 times the DC resistance. The rotor copper loss is calculated by subtracting the stator copper loss from the total measured loss or the rotor I^2R loss. The friction and windage loss may be assumed constant, irrespective of the load.

Efficiency = Rotor output/stator input

Output = Input − Losses

Example 4.1

Consider a three-phase 440 V, 50 Hz, six-pole induction motor. The motor takes 50 kW at 960 rpm for a certain load. Assume stator losses of 1 kW and friction and windage loss of 1.5 kW. To determine the percentage slip, rotor copper loss, rotor output, and efficiency of the motor, perform the following function:

Percentage slip

The synchronous speed of the motor = (50 × 120) / 6 = 6000 / 6 = 1000 rpm
Slip = (Synchronous speed − Actual speed) = 1000 − 960 = 40 rpm
Percentage slip = [(40 / 1000) × 100] = 4% = 0.04

Rotor copper loss

Rotor input   = 50 − 1 = 49 kW

Rotor copper loss = Rotor input × Slip

= 49 × 0.04 = 1.96 kW

Rotor output

Rotor output = Rotor input − Rotor copper loss − Friction and Windage loss

= 49 − 1.96 − 1.5

= 49 − 3.46

= 45.54 kW

Motor efficiency

Motor efficiency = Rotor output/Motor input

= 45.54 / 50 = 0.9108

= 91.08%

4.2.5   Principle and operation of a three-phase synchronous motor

The three-phase synchronous motor is considered as a constant-speed motor with large size and high ratings. As the name suggests, it runs from no load to full load at the same speed. As we have seen in an induction motor, there is a slip. However, here the motor runs at the same speed as that of the rotating magnetic field.

The construction of a synchronous motor is similar to that of an alternator. The stator winding is connected to the three-phase supply and the rotor has a DC field winding.

When three-phase voltage is supplied to the stator winding, this produces a rotating magnetic field. The torque .produced on the rotor is in a direction which will make the rotor field align with that of the stator. In a stationary rotor the torque is first in one direction and then the direction reverses depending on the relative positions of the stator's rotating magnetic field and the rotor's magnetic field. Due to the inertia of the rotor, it will not move in any direction. That is why synchronous motors are not self-starting.

Now, if it is made to run at some speed, then gradually, the stator and rotor poles of opposite polarity will be locked with each other causing the rotor to run in synchronism with that of the stator's rotating magnetic field. Thus, the rotor will run at a synchronous speed.

To make this happen, a synchronous motor is made to run like a normal induction motor at first, and then, like a synchronous motor.

For that purpose, the rotor has two windings, one is the AC winding, like a squirrel-cage winding or a wound rotor type, and the second is the DC winding. Stator winding is similar to the induction motor.

The three-phase synchronous motor differs from an induction motor, in that the rotor is wound and is connected to a DC source through the slip rings.

The motor starts as a normal induction motor (squirrel-cage/rotor-wound), once the speed of the rotor reaches 90–95% of synchronous speed; then a DC source is applied to the DC winding of the rotor. This in turn, produces the north and south poles in the rotor. The rotor magnet is now locked on to the rotating magnetic field of the stator and runs at a synchronous speed given by:

$$S = \frac{120\,f}{\dfrac{P}{3}}$$

Where

S = Synchronous speed in rpm
f = Frequency of AC source in Hz
P = Number of stator poles per phase.

However, it is necessary to have an angle between the centerline of the stator pole and the centerline of the rotor pole or field, as shown in Figure 4.3. If the excitation is kept constant, during the operation of the synchronous motor, and the load is increased, it produces a change in the current and in the power factor of the motor.

The characteristics of a three-phase synchronous motor are given below:

- Constant-speed motor
- Can be used to correct power factor of a three-phase system
- Is not self-starting
- Used generally in load applications requiring constant speed with infrequent starts and stops.

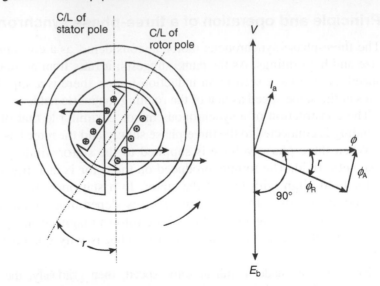

Figure 4.3
Synchronous motor action

4.2.6   'V' curves

If the armature current of a synchronous motor is plotted against the field current of the motor for a constant mechanical output, the curve produced is known as the V curve.

V curves can be drawn or obtained by tests, for various values of the loads maintained constant during these tests. Typical V curves for a synchronous motor for 'no load', 50% of full load, and full load are shown in Figure 4.4(a).

If a constant power is delivered, the armature current is reduced and the power factor lags as the field current increases. Then it becomes minimum at unity power factor. Again, it increases for leading power factor forming a V-shape characteristic.

If the power factor for constant output, is plotted against the field current, the plot will be an inversion of the V curve, as shown in Figure 4.4(b).

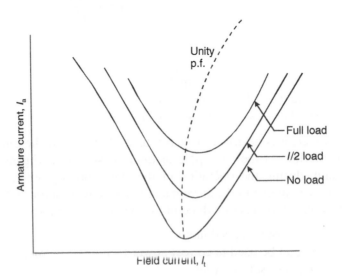

Figure 4.4(a)
Typical V curves for a synchronous motor

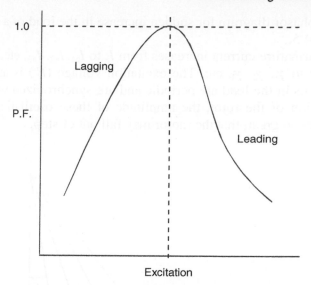

Figure 4.4(b)
Power factor v/s excitation

4.2.7 Losses and efficiency of a synchronous motor

The following are the types of losses in a synchronous motor:

- Fixed losses include core loss, friction and windage loss, and brush friction loss. These losses can be obtained from a no-load test.
- I^2R loss in armature windings and stray losses in conductors.
- Excitation circuit losses including field copper loss, rheostat loss, and brush contact losses.

The efficiency of a synchronous motor is determined as follows:

$$\text{Efficiency} = \frac{V_1 \cos \phi}{(V_1 \cos \phi \times 1.732 \times \text{Losses})}$$

4.2.8 Hunting of synchronous motors

This is a problem associated with synchronous motors. As the load on the motor increases, the angle between the stator pole and the locked rotor pole gradually increases. The rotor of the motor falls back by a certain angle, behind the poles of the forward rotating field, in order to produce the necessary torque. The stator current will also increase. While the motor speed slows down, it will remain synchronized, unless the load causes synchronization or the locking (of the rotor pole) breaks.

An increase in phase difference causes the motor to draw more current from the mains and increase power flow in the armature. Some of the kinetic energy of the rotating parts is transferred to the load during a speed slow down.

The motor cannot decelerate exactly at the required torque angle of the increased load. It passes beyond this, develops more torque, and increases the speed. This is followed by a reduction in speed and the cycle is repeated. If the load is suddenly thrown off, the rotor poles are pulled into almost opposition to the poles of the forward field, but due to the rotor inertia, the rotor poles travel too far, and are pulled back again. This results in oscillations about the position of equilibrium corresponding to the load conditions on the motor. This periodic change in speed is known as hunting.

The phasor diagram for sudden increase in the load of a synchronous motor is shown in Figure 4.5.

The armature current increases from I_a to I_{a1}, I_{a2}, I_{a3}, etc., and the torque angle increases from γ to $\gamma_1, \gamma_2, \gamma_3$, etc. The excitation voltage (E) is assumed to be a constant. If the variations in the load are periodic and are synchronized with the natural frequency of the oscillation of the rotor, the amplitude of these oscillations increases cumulatively, and becomes so great, that the motor may fall out of step.

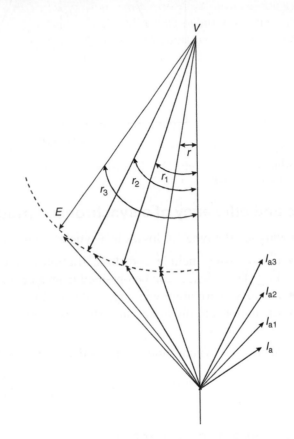

Figure 4.5
Current response for different loads with constant excitation for a synchronous motor

Hunting is also known as phase swinging.

In order to sort it out, it is required to damp the oscillations and prevent an increase in the amplitude of swinging.

This is achieved by providing a damper winding in the pole shoe of the synchronous motor. When hunting occurs, there is a shift of flux across the faces of the pole shoe, due to the effect of the armature reaction on the field flux. The shifting flux induces circulating currents in the damper winding. The kinetic energy of oscillations is dampened by being converted into heat energy. The induced current opposes the change in the relative positions of the armature flux and the field flux, and thus acts as an effective damper.

Thus, this problem of hunting in synchronous motors can be avoided.

4.2.9   Principle and operation of a three-phase wound rotor motor

In addition to a squirrel-cage, a wound-rotor induction motor has a series of coils set into the rotor. These are connected through the slip rings to external variable resistors.

The three-phase rotor connections are then brought up to slip rings. It is now possible to vary the rotor resistance, by introducing a different resistance in the rotor circuit through the slip rings provided. The speed and the starting torque will vary. The stator winding is similar to an induction motor with an individual three-phase winding placed 120° electrically apart.

In Figure 4.6, a wound-rotor induction motor is shown. It works as a normal induction motor with a three-phase supply given to stator windings. The only difference is that the speed varies depending on the rotor resistance.

High-starting torque is obtained with a low-starting current. When no resistance is introduced, the motor will run at full speed. As resistance is increased in the rotor circuit, the speed will reduce. For normal running, the slip rings are short-circuited.

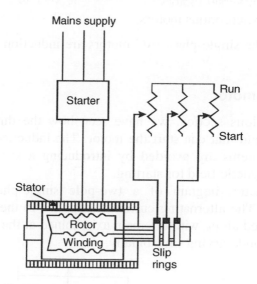

Figure 4.6
Wound-rotor induction motor

4.2.10 Wound-rotor motor startup

For a high-starting torque, the motor is started with the rotor circuit starter resistance in the circuit.

As the motor gains speed, the rotor resistance is reduced gradually. This will shift the synchronous speed and maximize the torque curve from the wound-rotor motor to an induction motor curve. Finally, the rotor or slip rings will be short-circuited.

For some motors, the rotor resistance is introduced in small steps.

Merely interchanging the two supply voltage leads can change the direction of the motor.

The following are the characteristics of a three-phase wound-rotor motor:

- Achieve high-starting torque with low-starting current
- Generally used in applications where on load starting is required
- Is self-starting
- Speed adjustment is possible up to a good extent
- Speed varies a lot when used with the rotor resistance in picture.

The high starting torque of the wound rotor motor and the capability to control the speed by varying the resistance has made this form of motor popular for lifting applications such as hoists and cranes. Also, the relatively lower starting current and the high torque makes it a popular choice for large capacity drives on weak electrical systems and for high inertia loads such as rotary kilns and blowers.

4.3   Fundamental of single-phase AC motors

4.3.1   Types of single-phase AC motors

Single-phase AC motors are also very widely used in industrial, commercial, as well as residential usage.

The main types of AC single-phase motors are:

- Induction motors
- Universal motors
- Synchronous motors.

Most of the single-phase AC motors are induction motors with different arrangements for starting.

Induction motor

In the previous section, we have seen how the three-phase supply sets up a rotating magnetic field that can start the motor. The induction motor, however, is a single-phase motor. Problems are avoided by introducing a second phase artificially to produce a rotating magnetic field for starting.

A schematic diagram of a two-pole single-phase induction motor is shown in Figure 4.7. The alternating current is supplied to the stator winding, the stator field axis remains fixed along with the main axis, joined at the centers of the two poles, alternating in polarity and varying sinusoidally.

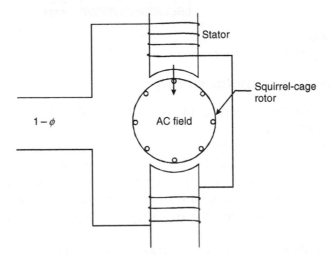

Figure 4.7
A single-phase induction motor with two poles

It can be shown mathematically that a pulsating magnetic field can be resolved into two equal rotating magnetic fields revolving in opposite directions, each trying to drag the rotor with it. Therefore the rotor remains stationary. However once the rotor starts rotating in any direction, the torque developed in that direction becomes higher and the motor continues to pick up speed till it reaches its rated speed. Thus a single phase induction motor is not self starting but requires some external means of initiating rotation.

This is done by having two stator windings. In addition, to produce a phase difference between two windings currents, a capacitor or an inductance is introduced in the starter-winding circuit. Once the motor reaches a normal speed, the starting winding is switched shut.

Once the motor starts, it will continue to run in the direction it has started. To reverse the direction of the motor any one of the winding leads has to be reversed.

Characteristics of single-phase induction motor:

- Works on a single-phase supply
- Auxiliary means of starting is required
- Are larger for the same horsepower compared with a three-phase motor
- Torque produced by motor is pulsating and irregular
- Mostly used in domestic and residential application.

4.3.2 Types of single-phase induction motors

Split-phase induction motor

This is a single-phase split in two-phase currents for creating a two-phase condition.

Thus, phase splitting creates a rotating field. In the split-phase motor, an auxiliary winding in the stator is used for starting, with either a resistance connected in series with the auxiliary winding, known as the resistance start, or a reactor in series with the main winding, known as the reactor start.

The start winding of a split motor is made up of a few turns of a small diameter wire giving a high resistance and a low reactance.

They are used in small machines that require low torque.

Capacitor-start induction motor

A single-phase induction motor with only a squirrel-cage rotor has no starting torque.

In this motor, a capacitor of suitable size is introduced in the starting winding circuit and a phase difference is created between the two windings.

In the capacitor-start single-phase motor, an auxiliary winding in the stator is connected in series with a capacitor and a centrifugal switch. During the starting and the accelerating period, the motor operates as a two-phase induction motor.

At about 67% of the full load speed, the switch disconnects the auxiliary circuit and the motor thereafter runs as a single-phase induction motor.

Single-value capacitor or capacitor split-phase motor

In a single-value capacitor, or a capacitor split-phase motor, a relatively small, continuously rated capacitor is permanently connected in one of the two-stator windings. This ensures that the motor starts and runs like a two-phase motor. The capacitor and the winding must be continuously rated. These motors are generally used for small pumps or such similar applications.

Capacitor-start/capacitor-run motor

In the capacitor-start, capacitor-run motor, the starting or the auxiliary winding remains connected in the circuit during running.

In order to have a high-starting torque, the capacitor rating should be high, but during running, a small value capacitor is suitable.

To solve this issue, two capacitors of different ratings are used. One is known as the starting capacitor and the other, as the running capacitor.

The starting capacitor is switched out of circuit once the motor reaches full speed. These motors have good starting torque and a quiet running.

Shaded-pole induction motor

This is another method to create two phases, with phase differences using pole shading. The shaded pole has salient poles, each pole split into two sections. A small part of each pole face is wound with a short-circuited copper ring.

When supply is given, the flux under the shaded portion lags behind that of the un-shaded portion. Thus a phase difference is created. This causes the creation of a rotating magnetic field.

The shaded pole motor is inexpensive, since it does not consist of an auxiliary winding or other mechanisms. These are low-torque motors, generally used for small fans and blowers.

Repulsion-start single-phase motor

The principle of operation of the repulsion-type single-phase motor is an interesting contrast with other motors.

In a repulsion-start single-phase motor, a drum-wound rotor is used, which is similar to a squirrel-cage rotor. The circuit is connected to a commutator with a pair of short-circuited brushes. These are set such that, the magnetic axis of the rotor winding is inclined to the magnetic axis of the stator winding. The current flowing in the rotor circuit reacts with the field, to produce a starting and an accelerating torque. At about 67% of the full load speed, the brushes are lifted, the commutator is short-circuited and the motor runs as a single-phase squirrel-cage motor.

The repulsion induction motor employs a repulsion winding on the rotor during the start and running. The repulsion induction motor has an inner squirrel-cage winding and an outer winding on the rotor, which acts as a repulsion winding. As the motor speeds up, the induced rotor current partially shifts, from the repulsion winding to the squirrel-cage winding and the motor runs partly as an induction motor.

Repulsion-start motors have a high-starting torque. Changing the position of the brushes can vary their speed.

Series-wound single-phase or universal motor

The series-wound single-phase motor, has a rotor winding in series with the stator winding, similar to the series-wound DC motor.

This is similar to a DC series motor with a similar high-starting torque. This motor is also called the universal motor because it is designed to operate on either AC or DC supply.

When an AC is applied, there is an instantaneous change in the field and armature polarities. Since the field winding has a low inductance, it creates reversals in the current direction at the proper timings.

As the motor can also be operated on direct current (DC), it is also known as the universal motor.

Single-phase synchronous motor

This is similar to a three-phase synchronous motor in terms of speed, as this also runs at a constant speed. The stator winding is connected to a single-phase supply, while the rotor is made up of a permanent magnetic material.

The resistor and the capacitor are connected in series with one winding. This results in a 90° phase shift between the two windings.

For the first 90°, one set of the windings produces a force that attracts the electromagnet rotor. The second set of phase windings again attracts a new position. This provides the rotor, the force required to start and continue running.

4.4 DC motors

The development of the DC motor preceded that of the AC motor. Today, even with the preponderance of AC motor applications, notably in traction, the DC motor retains its position due to its unique characteristics.

DC motors are used for applications that require a wide range of torque and good speed control.

4.4.1 Types of DC motors

The basic working principle of DC motors has already been dealt with. This section details the types of DC motors. The different types depend on the method of connection of the armature and the field winding and they are classified as below:

- Separately excited motor
- Series DC motors
- Shunt DC motors
- Compound DC motors.

Separately excited motor

In a separately excited motor, as shown in Figure 4.8, the field supply is provided in a manner other than by an armature supply.

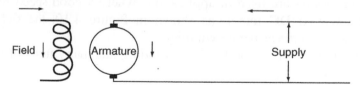

Figure 4.8
Separately excited DC motor

Since the field current is taken from a separate supply, field current is independent of the load.

The flux remains essentially constant and does not affect either the speed or the torque.

These types of motors are not used generally, as they require a separate field supply and also large variations in field current apart from the armature circuit.

Series DC motor

As shown in Figure 4.9, the motor has the armature and field winding connected in series, therefore, its name – series motor. Field winding is made up of a relatively few turns of a large diameter wire so that the field resistance remains low.

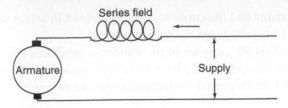

Figure 4.9
Series DC motor

In the series-wound DC motor, the field windings are fixed in the stator frame and the armature windings are placed around the rotor. These are connected in series. The current passing through the armature also passes through the field. In a series-wound motor, any increase in the load results in more current passing through the armature and the field windings.

As the increased current strengthens the field, the motor speed decreases.

Conversely, if the load is decreased, the field is weakened and the speed increases. For lighter loads, the speed increase may be excessive and undesirable.

In order to control this, the series-wound DC motors are usually directly connected or geared to the load to prevent runaway.

At times, a series-wound motor designated as a series-shunt wound, is provided with a light shunt field winding, to prevent the dangerously high speeds at light loads. The increase in armature current with the increasing load produces an increased torque, so that the series-wound motor is suited to heavy staring duty. The motor speed can be adjusted with a variable resistance placed in series with the motor, but due to the variation with the load, the speed cannot be maintained at a constant. The series-wound DC motors are used for hoists, cranes, and elevators.

Shunt DC motors

Shunt motors are used in applications where a good speed regulation is required. In the shunt-wound DC motor, as shown in Figure 4.10, the field winding is connected in parallel with the armature winding.

In this type of motor, the field winding has many turns of small diameter wire to keep the resistance high.

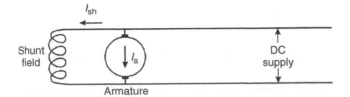

Figure 4.10
Shunt DC motor

The strength of the field is not affected appreciably due to changes in the load. A more or less constant speed is obtainable.

Shunt-wound DC motors are used where a more or less constant speed, a low staring torque, and a light overload on the motor are required.

The shunt-wound motor can also work as an adjustable-speed motor by means of the armature control or field control.

If a variable resistance is placed in the field circuit, the amount of current in the field windings can be controlled and the speed of the motor can be controlled.

As the motor speed increases, the torque decreases proportionately, resulting in an approximately constant horsepower.

If a variable resistance is placed in the armature circuit, the voltage applied to the armature can be reduced, and hence the motor speed can be reduced. With an armature control, speed regulation becomes poorer as the speed is decreased. Since the current in the field remains unchanged, the torque remains constant.

Adjustable-speed shunt-wound motors are used on large machines for boring mills, lathes, planners, etc. These are particularly adapted to spindle drive because of the constant horse-power characteristic that permits heavy cuts at low speeds and light cuts at high speeds.

Adjustable-voltage shunt-wound motor drive

Due to the requirement of DC power, the application of shunt-wound DC motors has been limited. Extensive use of the shunt-wound motors has been made possible by a combination drive that includes a means of converting AC to DC. A self-contained unit may achieve the conversion of AC to DC. This consists of a separately excited DC generator driven by a constant-speed AC motor, connected to the regular AC supply. An electronic rectifier with suitable controls, connected to the regular AC supply, can also achieve this. The conversion of AC to DC with an electronic rectifier has the advantage of causing no vibrations.

In an adjustable-speed, shunt-wound motor drive, speed control is affected by varying the voltage applied to the armature while supplying a constant voltage to the field.

In addition to providing for the adjustment of the voltage supplied by the converter, to the armature of the shunt-wound motor, the amount of current passing through the motor field may also be controlled. In fact, a single control may be provided to vary the motor speed, from a minimum base speed, by varying the current flowing through the field. With such control, the motor operates at a constant torque up to the base speed and at constant horsepower above the base speed.

Adjustable-speed shunt-wound motor drives are also called DC-adjustable voltage drives. These drives are used for milling machines, boring mills, lathes, and other industrial applications where wide, step-less speed control, uniform speed under all operating conditions, a constant torque acceleration and adaptability to automatic operational control are required.

Compound-wound motors

These types of motors provide desirable characteristics of both series- and shunt-wound motors. They provide high-starting torque as in a series motor, as well as good speed regulation as in a shunt-wound motor.

In these types of motors, there are two field windings; one is in series with the armature, while the other is in parallel with the armature.

In the compound-wound motor, the speed variation due to load changes is much less than that in the series-wound motor, but greater than that in the shunt-wound motor. The compound-wound motor has greater starting torque than the shunt-wound motor and is able to withstand heavier loads for narrow adjustable-speed range (Figures 4.11 and 4.12).

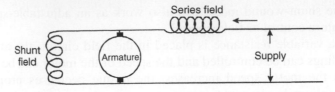

Figure 4.11
Compound motor short shunt connected

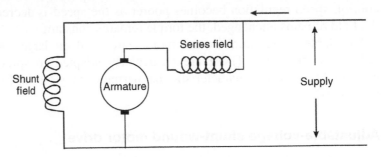

Figure 4.12
Compound motor long shunt connected

4.4.2   Characteristics of DC motors

- DC motors have good starting torque as well as good speed-regulation capabilities.
- Permanent magnet-type DC motors are used for exact positioning of objects with high-operating torques.
- Series type gives high-starting torque; hence the ability to start with high loads.
- Series motors when operated with no loads can attain high speeds causing harm to motor.
- Shunt-type motors have good speed regulation.
- Direction reversal of DC motor can be done by changing the leads of the armature or the field.
- Speed of DC motor changes either by changing armature voltage or field current.
- If armature voltage is increased, speed increases till base speed and vice versa.
- Similarly, speed can be increased above base speed, by decreasing field current.
- In shunt motors torque is proportional to armature current.
- In series motors torque is proportional to square of armature current.

Example 4.2

Consider a 250 V DC shunt motor with armature resistance of 0.5 Ω and a shunt field resistance of 125 Ω. When it is running light the motor current taken is 5 A. The efficiency of the motor when taking current of 52 A from the supply is calculated as follows:

On no load:

$$I_t = \frac{V}{R_f}$$

$$= \frac{250}{125}$$

$$= 2\,A$$

So

$$I_o = I - I_t$$
$$= 5 - 2$$
$$= 3\,\text{A}$$

Iron and mechanical loss = $V \times I_o = 250 \times 3 = 750$ W
Field copper loss = $I^2 R_f = 2 \times 2 \times 125 = 500$ W
Total loss = $750 + 500 = 1250$ W

On load:

Input = $V \times I = 250 \times 52 = 13\,000$ W
$I_a = 52 - 2 = 50$ A

Armature copper loss = $I_a^2 R_a = (50)^2 \times 0.5 = 1250$ W
Total losses = $1250 + 1250 = 2500$ W
Efficiency = (Input – losses) / Input
$\qquad\qquad = (13 - 2.5)/13 = 0.808 = 80.8\%$

4.5 Motor enclosures

Motors are manufactured in standard frame sizes, corresponding to the rated output. The size of the frame naturally increases with the rated output. The frame size standards stipulate various dimensions like the shaft center height, axial distance between shaft end and the nearest pair of mounting holes, axial distance between the sets of mounting holes and other mounting dimensions. This ensures interchangeability between motors of the same frame size manufactured by different vendors. In a given frame size, a number of designs are available to suit various applications, for example, types of mounting like foot-mounting or flange-mounting.

The type of enclosure required, depends upon the conditions under which the motor has to work. It is therefore selected so as to protect the internal parts against the ingress of dust and water. At the same time the enclosure is expected to protect the surrounding areas from the internal, live, and moving parts of the motor.

The type of insulation and the type of cooling required is selected, depending on the temperature rise and the operating temperature limits.

The position of the terminal box, which is either on top or on the side, is often required to be specified while ordering a motor.

The frame itself is available in a number of designs to suit the requirements of site conditions, duty, etc. Some of the standard designs are as follows:

1. *Totally enclosed, non-ventilated type*

Such motors are limited to sizes up to 2 or 3 kW. The cooling is by surface radiation as there are no openings for ventilation. The frame is of solid construction.

2. *Splash-proof type*

The frames of such motors, incorporate ventilated openings, so constructed that the liquid drops and dust particles that fall vertically or greater than 10° from the vertical angle, cannot enter the motor directly. They also cannot enter the motor by striking or running along the outer surface.

This is the extension of drip-proof motors and is called the hose-proof type.

3. *Totally enclosed fan-cooled*

Totally enclosed fan-cooled motors, incorporate a fan, mounted on the motor shaft. The fan draws air and forces it between the inner fully enclosed frame and an outer shell. An internal fan carries the heat that is generated internally to the enclosed frame, which is cooled by the air drawn by the fan on the shaft. The enclosing frame protects the motor against corrosive and abrasive effects of dust, moisture, etc.

4. *Protected-type*

Protected-type motors contain feature-perforated covers for the openings in the end shields. Thus the internal and live parts of the motor are mechanically protected using wire mesh or metal covers without affecting the flow of air.

5. *Drip-proof type*

The frames in such motors afford protection against liquid drops or dust particles falling on the machine at angles greater than 15° from the vertical. Water drops and particles cannot enter the motor directly or indirectly by either striking or running along the surface.

The ventilated openings are protected by use of a hood.

6. *Pipe or duct ventilation type*

Sometimes the air surrounding the motor is such that, if it is passed through the motor winding, it can damage it.

In such cases, clean air can be brought from outside, and by means of a pipe or a duct, it can be used for motor ventilation. For force induction of air, a blower is installed at either the entry side or the exit side of the duct.

4.5.1   Motor nameplate

The motor nameplate gives important information about the motor. It gives among others, information about the following:

- Motor rating
- Motor supply details
- Motor connection details
- Motor frame type and size
- Motor rpm
- Permissible temperature rise
- Motor duty
- Enclosure type
- Number of poles.

The nameplate gives details about the motor at a glance.

4.6   Motor terminal Identification and connection diagram

Usually, the motor terminal connection diagram is given on the motor. For a three-phase motor, three winding end connections are shown – U1 and U2, V1 and V2, and W1 and W2.

Motor terminals and connection diagrams for Autotransformer or DOL starting with three- or six- leads connection are shown in Figure 4.13(a).

Terminal connection diagram

DOL or autotransformer starting with three or six leads.
Number of leads connection

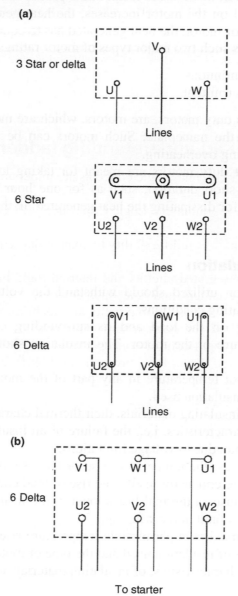

(a)

3 Star or delta

Lines

6 Star

Lines

6 Delta

Lines

(b)

6 Delta

To starter

Figure 4.13
Typical circuit diagram (a) DOL starter; (b) Delta starter

Motor terminals and connection diagrams for star–delta starting with six-leads connection is shown in Figure 4.13(b).

Similarly, for a DC motor, field connections are indicated as F1 and F2 while armature winding is shown as A1 and A2. If the DC motor is of a series-type or a compound type, then the series winding leads are shown as S1 and S2.

4.7 Motor rating and insulation types

4.7.1 Motor rating

Motor rating is defined, as the output of a motor under prescribed working conditions, with a temperature rise below specified limits.

Motors suffer various losses like core loss, stator loss, rotor loss, winding loss, and friction loss. All these losses result in the production of heat.

As the load on the motor increases, the heat generated also increases. To maintain a healthy state of motor, the heat generated has to equal the heat dissipated.

There are as such two major types of motor ratings:

- Continuous
- Intermittent.

Continuous duty motors, are motors, which are meant to give the continuous rated load specified on the nameplate. Such motors can be operated at these load continuously, without causing overheating.

Intermittent duty motors are meant for taking loads above the maximum continuous rating, for a short duration, such as for one hour or so. This allows the motor, to get enough time for dissipating the heat generated, in the time intervals when the motor is not running.

4.7.2 Motor insulation

The insulation utilized should withstand the voltage fluctuations of the motor under varying operating conditions.

Depending on the load and its surrounding conditions, there could be a rise in the temperature of the motor. The insulation should withstand such temperature rises also.

The hot-spot temperature in any part of the motor should not exceed the permissible limit of the insulation used.

In case of insulating materials, their thermal characteristics are more sensitive than their dielectric characteristics, i.e., the failure of an insulating material is more due to thermal limitations than due to voltage limitations.

In most cases, the temperature rise or the rise in load does not produce a fault in the winding of the conductor itself. The rise of load current or greater fault current, when it is excessive, causes a thermal breakdown in the insulation covering the conductor. This is what creates a fault in the winding.

Thus, the maximum permissible temperature rise, in electrical motors, must be in tune with the type of insulation used and the type of motor.

The main characteristics, of insulating materials used in electrical machines are:

- Dielectric strength
- Thermal strength.

The insulating material used for the electrical machines should satisfy the following requirements:

- High dielectric strength, high specific resistance, and minimum loss in alternating electric field
- High mechanical strength and elasticity of material

- Thermal strength of insulation; the insulating material should preserve its insulation and mechanical properties when subjected to the operating temperatures of the windings for a long time
- The material should remain unaffected by chemical influences.

The temperature rise permissible can be determined, by deducting the ambient temperature, from the maximum permissible temperature.

For electrical machines, the following, are the types of insulating material that have been classified and standardized as follows:

- *Class A insulation*: Cotton, silk, paper, and similar organic materials, impregnated or immersed in oil, and enamel applied on enameled wires. The limiting hot-spot temperature for Class A insulation is 105 °C.
- *Class E insulation*: An intermediate class of insulating materials between Class A and Class B insulation materials.
- *Class B insulation*: Mica, asbestos, glass fiber, and similar inorganic materials, in built-up form with organic binding substances. The limiting hot-spot temperature for Class B Insulation is 130 °C.
- *Class F insulation*: Includes insulation having mica, asbestos, or glass fiber base with a silicone or a similar high-temperature-resistant binding material. The limiting hot-spot temperature for Class F insulation is 155 °C.
- *Class H insulation*: Includes insulation having mica, asbestos, or glass fiber base with a silicone or a similar high-temperature-resistant binding material. The limiting hot-spot temperature for Class H insulation is 180 °C.

4.7.3   AC motor connections

(a) *Multispeed motor*

A three-phase induction motor is a constant-speed machine. The speed control of induction motors can be achieved by:

- Changing the applied voltage
- Changing the applied frequency
- Changing the number of poles.

The first two methods, however, are rarely used because of the problems associated with reducing the voltage and frequency. The last method is well suited for squirrel-cage motors, as the squirrel-cage rotor adapts itself to any reasonable number of stator poles. The change in the number of stator poles is achieved by providing two or more entirely independent stator windings. Each winding gives a different number of poles and hence a different speed.

Each of the windings is terminated on a different set of terminals, which can be connected up and switched, to connect the winding to the supply. Only one winding is used at a time, the other, being entirely disconnected. This method finds application in elevators, traction, and small motor-driven machine tools.

(b) *Dual-voltage motor*

A dual-voltage motor is a single-phase induction motor. It can be operated from two AC voltages, either 110 or 220 V. Such motors, have two main windings and one starting winding. A suitable number of leads are brought out to permit changeover from one voltage to another.

When the motor is to operate on a lower voltage, the two main windings are connected in parallel. On higher voltage, they are connected in series. The starting winding is always operated on the low voltage mode, for which purpose it is connected across one of the main windings.

4.7.4 DC motor connection

DC motors, may be of a series, shunt, or compound type. Depending on the type of motor, it will have field-winding, armature-winding, or series-winding terminal connections.

4.8 Operating a motor for forward and reverse operation

4.8.1 Induction motor

In certain applications like lifts, cranes, etc., it is required that the motor be capable of reversing. This requires a reversible motor, i.e., one in which the shaft has a bi-directional rotation, both clockwise and anti-clockwise. In the induction motor, this is quite simply achieved by changing over any two of the leads.

In Figure 4.14, the direction of an induction motor with DOL starter is reversed. Phases L1 and L2 are interchanged; you can always change direction of the induction motor by interchanging any two phases. In larger ratings of motors the cooling fans may be suitable for one direction of rotation only. If bi-directional operation is anticipated, the same should be specifically stated to the manufacturer while ordering.

This can be achieved by changing the leads, either at the motor terminals or at the starter end connection. It is always advisable, to make changes to the lines going to the starter, as there may be some confusion in case of motors arranged for star–delta starting while making changes at the motor terminals.

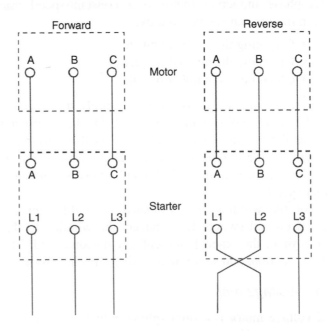

Figure 4.14
Reversing a three-phase motor

In Figure 4.15, reversing connections of a three-phase induction motor with star–delta starter are shown. Phases L1 and L2 are interchanged; you can always change direction of the induction motor by interchanging any two phases.

As discussed earlier, it is advisable to make changes in the lines going to the starter. There may be some confusion, in the case of motors arranged for star–delta starting, while making changes at the motor terminals.

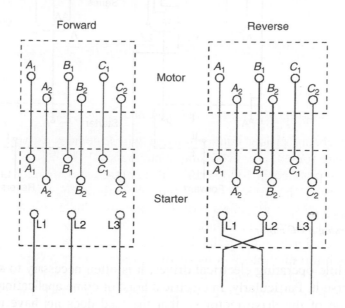

Figure 4.15
Reversing a star/delta motor

4.8.2   DC motors

In case of a DC series motor, the direction of the motor can be reversed either by changing the direction of the armature current or the field current.

In the same way for a shunt motor, the direction of the motor can be changed either by changing the direction of the armature current or field current. In the case of a compound motor, the direction of rotation can be reversed by changing the direction of the armature current or of both field currents, i.e., series and shunt field windings.

The industrial standard, however, is to change the armature current direction, keeping the direction of current in the field unchanged.

This is shown in Figure 4.16. At the motor terminal box, the starter cable is changed from A to AA, and the interconnecting link is changed from Y to AA. Any type of DC machine can be reversed by reversing the current flow in the armature and interpoles. It should be noted that the brush position might need to be changed to suit the new rotation.

4.9   Motor braking methods

Motor braking is to stop a running motor.

Removal of the supply given to the motor will make it stop. However, due to inertia, the motor will tend to rotate for some time before coming to a complete halt.

To stop the motor quickly, a braking mechanism is required. This is called motor braking.

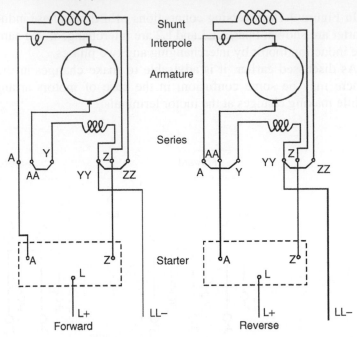

Figure 4.16
Reversing a DC motor

While operating electrical drives, it is often necessary to stop the motor quickly and to reverse it. Particularly, in electrical hoist or crane applications, it is required to control the torque of the drive motor so that the load does not have undesired acceleration. Some applications require accurate positioning of the motor shaft.

The speed and accuracy of stopping and reversing operations of motors improve the productivity of the system. For example, take rolling mill motor application.

In such applications, braking torque is required which may be either electrical or mechanical. Braking can be broadly classified as:

- Electrical braking
- Mechanical braking.

(a) *Electrical braking*

In electrical braking, the winding of the motor is used to produce a braking torque. A braking torque is developed during the braking operation. This braking torque opposes the motion of the rotating member or shaft. This is achieved by suitably changing the electrical connections of the motor. The motor operates on a speed–torque characteristic, depending upon the method of braking employed.

(b) *Mechanical braking*

In mechanical braking, the frictional force between the rotating parts and the brake drums produces the required braking. To achieve this, mechanical equipment, such as brake linings and brake drums, are required. Whether electrical or mechanical, the braking of the drive should stop the motor at the specified point of time and position.

A brief comparison of electrical and mechanical braking is given in Table 4.1.

	Electrical Braking	Mechanical Braking
1. Maintenance	Little maintenance. Dust-free operation due to absence of mechanical equipment	Mechanical brakes require frequent maintenance. Adjustment of brakes and replacement of brake linings due to wear and tear
2. Energy utilization	Energy of rotating parts can be converted to electrical energy. This can be utilized or returned to the mains during braking	Energy of rotating parts is wasted as heat in friction. Heat is generated during braking
3. Smoothness of braking	Braking is smooth, without snatching	Depending on conditions, braking may not be smooth
4. Equipment	Equipment of higher rating than the motor rating may be required in certain braking applications	Equipment like brake shoes, brake lining, brake drum are required
5. Holding	To producing holding torque, electrical energy is required for operation	Mechanical braking can be applied to hold the rotating part or shaft at a particular position

Table 4.1
Comparison of electrical and mechanical braking

Conclusion

From the above comparison of electrical and mechanical brakings, it is seen that electrical braking is more effective and superior to mechanical braking. However, for safety reasons, in hoist or crane applications a standby mechanical brake system is also provided to avoid accidents in case of power failure. However, in case of severe operating cycle of the motor, electrical braking should be employed only when it is highly desirable to control retardation and limit the braking time.

4.9.1   Electrical motors braking methods

Electrical braking may be achieved by the following:

- Counter current braking or plugging
- Regenerative braking
- Dynamic or rheostatic braking.

Electromechanical braking may be achieved through an electromechanical brake.

(a) *Counter current braking or plugging*

This is accomplished by momentarily connecting the motor in a forward direction, when the motor is already running in the reverse direction. It is accomplished in DC motors, by reversing the armature supply leads, so that the motor draws a current to develop a reverse torque, to oppose its already existing rotation.

The motor acts as a brake and comes quickly to rest but has the tendency to accelerate in a reverse direction. If the reversal is not required, the supply to the motor should be cut off at zero speed.

In case of a DC motor, this is achieved by reversing the polarity of the supply voltage to the armature; however, in case of AC motors the phase sequence is interchanged.

This method of braking is also used to maintain a constant speed when the load tries to accelerate the rotor to high speeds.

This method is inefficient because of the power loss in resistors, used for limiting the current due to an interconnection. The mechanical energy is converted to heat and additional power is required. At the same time, if for a large motor, sudden torque is applied, in the reverse direction by plugging, then it may result in damaging the machinery. This may result in a high current flow through the system.

(b) *Electrical braking*

In electrical braking, it is possible to convert the kinetic energy of the rotating parts to electrical energy and return it back to the mains or dissipate it in a resistance.

(c) *Regenerative braking*

The braking is called regenerative when the energy is returned to the mains supply. This method uses the motor as a generator during braking, developing a retarding torque, which acts on the running motor and halts it. The kinetic energy and potential energy, minus the losses of the motor are returned to the mains and the motor runs at a constant speed. Thus, regenerative braking eliminates the tendency of the load to accelerate the motor.

(d) *Dynamic or rheostat braking*

The braking is called dynamic or rheostatic braking when the energy is dissipated as heat in a resistance. In either case, the machine operates as a generator. Electric machines are capable of smooth transition from motor to generator action. In dynamic braking, the motor must be switched to the load or braking resistor keeping the field constant.

(e) *Electromechanical friction braking*

This is an external brake, used for stopping the motor as well as holding the motor at a single position. It consists of a solenoid along with a drum brake arrangement. When the motor is running, the solenoid is energized; so, it keeps the brake shoes away from the rotor shaft. As the motor is turned off, the solenoid is de-energized and the brake shoe acts on the rotating shaft. There is a braking action due to the friction between the brake shoe and the rotating shaft.

This kind of brake holds the load at one position, even after the motor stops. It is therefore used in applications that require the motor to be held at one position with the load, like in crane applications.

These brakes require more maintenance than purely electrical braking, because of the wear and tear on the brake shoe mechanism.

4.9.2   Induction motors braking methods

1. *Regenerative braking*

In an induction motor, when the rotor runs faster than the stator field's synchronous speed, the slip becomes negative and the machine generates power.

Whenever the motor has a tendency to run faster than the rotating field, regenerative braking occurs and the kinetic energy of the rotating parts is returned to the mains.

The speed torque curve extends to the second quadrant as shown in Figure 4.17.

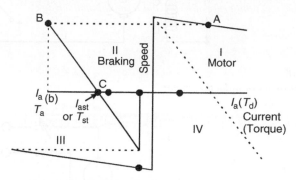

Figure 4.17
Regenerative braking of induction motor

As the speed of the motor decreases, the braking torque makes the motor run at a constant speed and arrests its tendency to rotate faster. Due to the effects of stator resistance, the maximum torque developed during regeneration is greater than the maximum torque during motoring.

For example, in cranes and hoists, the motor has a tendency to run faster than the synchronous speed. This situation can occur when a hoist is raising an empty cage. Due to counter weight, the cage may acquire dangerous speeds. The transition takes place almost automatically, a torque is developed to arrest the acceleration, and regeneration takes place. Automatic regeneration arrests any undue acceleration. In such cases, the rotor resistance control could be employed to get a better braking torque.

Regenerative braking is also possible with a pole change motor, when the speed is changed from high to low. It can also be easily accomplished, in variable frequency drives by decreasing the frequency of the motor momentarily – the synchronous speed decreases and conditions favorable to regeneration are created.

As the motor speed decreases, the frequency is continuously reduced so that the braking takes place at a constant torque and stator current, until the motor comes to zero speed. During regenerative braking, there is a possibility of dangerous speeds, if the operating point during the braking falls in the unstable region of the characteristic. This happens if the load torque is greater than the breakdown torque of the motor. The torque developed cannot break the motor and undue acceleration takes place. This possibility can be eliminated by means of a high resistance in the rotor.

2. *Dynamic braking*

Dynamic braking is used to brake non-reversing drives. The stator is transferred from AC mains to DC mains, as shown in Figure 4.18.

The DC flowing through the stator sets up a stationary field. This induces rotor currents that produce a torque, to bring the rotor to rest quickly. The developed torque and retardation during braking may be controlled by the amount of the DC power. Additional resistances r_1 and r_{2e} in the stator and rotor circuits control the DC excitation and braking torques, respectively.

An equivalent circuit and phasor diagram of the motor during dynamic braking are shown in Figure 4.19.

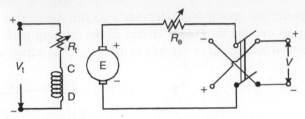

Figure 4.18
Dynamic braking of induction motor

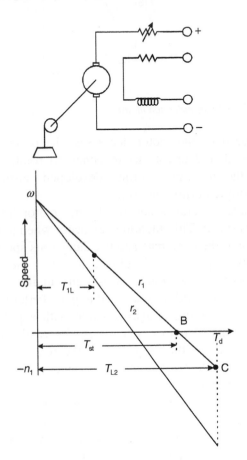

Figure 4.19
Equivalent circuit and phasor diagram of the motor during dynamic braking

When the stator is fed from the DC, the MMF produced is stationary. The MMF depends upon the stator connections for feeding the DC, the number of turns, and the current. The possible connections of the stator for feeding the DC are shown in Figure 4.20.

Dynamic braking is employed in conjunction with automatic control. Induction motors are more popular in hoists than in DC motors due to this reason. Methods of feeding the DC supply to the stator are shown in Figure 4.20. The limiting resistor R controls DC excitation.

Torque control is achieved by a rotor resistance variation. Alternately, an AC supply with a bridge rectifier may be used to feed the motor.

In AC dynamic braking, the stator is switched to a capacitance bank. The machine runs as a self-excited induction generator. All the mechanical energy is dissipated as electrical energy in the rotor resistance. This method is un-economical due to a high cost of capacitors.

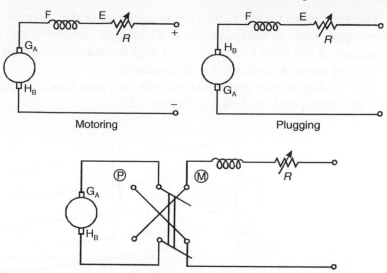

Reconnecting of DC Series motor for reverse current braking (Plugging)

Figure 4.20
Stator connections for DC dynamic braking

3. *Counter current braking or plugging*

In an induction motor, by changing the phase sequence of the input supply, the direction of the stator field can be reversed. This is also called plugging.

In practice, interchanging the supply to any two terminals of the motors, as shown in Figure 4.21, does this.

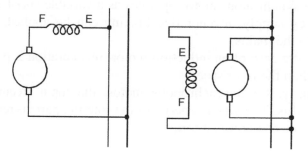

Figure 4.21
Changing of phase sequence for counter current braking of a three-phase induction motor

When the phase sequence of the input supply is changed, this reverses the direction of the revolving flux, which produces a torque in the reverse direction, thus applying a brake on the motor. During this braking action, the motor absorbs kinetic energy from the revolving load, causing speed reduction and bringing the motor to a rest. The motor must be switched off, as it approaches zero speed.

In high-capacity motors, if sudden torque is applied in a reverse direction (plugging), without slowing down the motor, then it can result in a mechanical damage. To avoid this, anti-plugging protection is used, which does not allow a reverse torque to be applied, unless the speed of the motor reduces below the acceptable value.

The speed–torque characteristics of an induction motor can be modified by varying the rotor resistance.

The maximum torque point can be achieved in the range of slips 1–2, where the torque developed tends to brake the rotor. The torque developed can also be used to arrest the tendency of the rotor to accelerate. A high resistance is introduced in the rotor, so that the operating point shifts to the fourth quadrant.

The braking torque developed prevents any acceleration of the rotor which now works at a uniform speed, as shown in Figure 4.22.

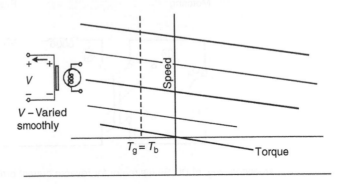

Figure 4.22
Speed-torque curve of an induction motor during plugging

The total braking torque (T_P) may be controlled with a variable rotor resistance, which limits the braking current.

4.9.3   Synchronous motors braking methods

(a) *Regenerative braking*

When a synchronous motor operates as a variable speed drive (VSD), utilizing a variable frequency supply, regenerative braking can be applied, and all the kinetic energy will return to the mains.

As in the case of an induction motor, regeneration is possible if the synchronous speed is less than the rotor speed.

Similarly, for a synchronous motor, the input frequency is gradually decreased to achieve this. The kinetic energy of the rotating parts is returned to the mains. The braking takes place at a constant torque.

(b) *Dynamic braking*

To achieve dynamic braking, a synchronous motor is switched on to a three-phase balanced-resistive load, after disconnecting it from the mains, keeping the excitation constant.

To achieve a greater braking torque for effective braking, the excitation may be increased. The terminal voltage and current decrease as the speed decreases. At very low speeds, the resistance effect is considerable. The resistance affects the speed at the maximum torque. The maximum torque can be ideally made to occur just before the motor is stopped.

(c) *Plugging*

The braking of a synchronous motor, by the plugging method, has major disadvantages, such as very heavy braking current flow. This causes line disturbances and an ineffective torque.

Plugging can be used for braking, if the motor is a synchronous induction motor and only if the machine is working as an induction motor.

4.9.4 DC motors

A motor and its load can be brought to rest quickly by using either of the following:

- Friction braking
- Electric braking.

The common mechanical brake has one drawback. Smooth braking is difficult to achieve, apart from the fact that, it largely depends on the operator's skill.

Electric braking methods eliminate the need for brake linings, levers, etc. There are three methods of electric braking both for shunt and series motors:

- Rheostatic or dynamic braking
- Plugging or reversal of torque
- Regenerative braking.

A friction brake, however, is necessary for holding the motor, even after it has been brought to rest. This is so even in the case of electric braking.

4.9.5 Shunt motor

(a) *Rheostatic or dynamic braking*

In this method, the armature of the motor is disconnected from the supply and is connected across a variable resistance. The field winding is not disturbed.

As shown in Figure 4.23, a braking resistor is placed across the armature winding with an NC contact of motor start contact (K) in series. So, when the motor runs, the braking resistor remains isolated.

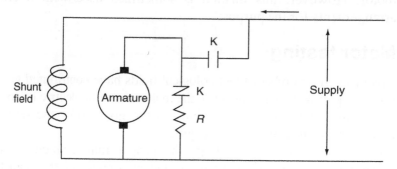

Figure 4.23
Dynamic braking of DC motor

When the stop button is pressed, the braking resistor will function as a load. It will dissipate the stored energy till the motor stops.

Varying the series resistance controls the braking. The smaller the value of the resistance, the higher the rate at which the energy stored will be dissipated and the earlier the motor will stop.

(b) *Plugging or reverse current braking*

In this method, the connections to the armature terminals are reversed so that the motor tends to run in the opposite direction.

Plugging gives a greater braking torque than rheostatic braking. Obviously, during plugging, the power is drawn from the supply and is dissipated in the form of heat. If the motor is of a large capacity, then due to plugging, there may be a rise in the current. The resultant application of sudden reverse torque may damage the machinery.

(c) *Regenerative braking*

This method is used when the load on the motor has an overhauling characteristic, as happens when lowering the load in a crane or in the downhill motion of an electric train. Regeneration takes place in such a situation, as the motor acts as a generator, returning power to the line, which may be used for supplying another train, thereby relieving the powerhouse of part of the load.

4.9.6   Series motor

(a) *Rheostatic braking*

In this type of braking, the motor is disconnected from the supply. The field connections are reversed and the motor is connected in series with a variable resistance. The machine now acts as a generator. The variable resistance employed for starting is, in itself, useful for braking, in a number of applications.

(b) *Plugging or reverse current braking*

As in the case of shunt motors, in the series motor too, the connections of the armature are reversed and a variable resistance is introduced in series with the armature.

(c) *Regenerative braking*

This type of braking of a series motor is not possible, without modification in the series motor. However, this method is sometimes used with traction motors, with special arrangements for the purpose.

4.10   Motor testing

Advances in a number of technological fields have contributed greatly to the development of testing equipment. This has made the maintenance of electric motors a more precise science.

The task of keeping all 'systems on the go' is now the responsibility of a 'cut to the bone' staff of technicians and engineers.

Despite all the hi-technology, the proper maintenance of equipment like motors is dependent on training, motivation, and awareness of the man behind the machine. The following tools and equipment are now available for monitoring, measuring, and metering the various parameters of the motors in a motor performance/operation.

Different tests provide various data, related to the motor, which give an idea about the probable performance of the motor and its efficiency.

4.10.1   Methods of testing DC machines

DC motors are tested by using the following methods:

1. *Direct loading or brake test*

This test is meant for calculating the efficiency of a motor and is a direct loading test. It is used for small motors only. The motor is loaded directly by a mechanical rope or a belt brake.

A typical arrangement for the direct loading of the motor for a brake test is shown in Figure 4.24.

In this method, a force is applied by adjusting the brake and is measured in Newtons using an attached spring balance.

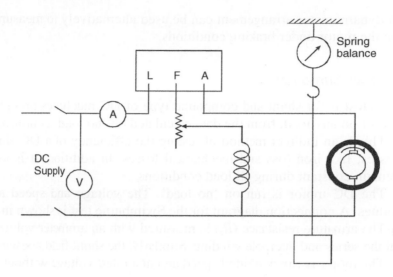

Figure 4.24
Test arrangement of direct loading of motor for brake test

Torque is then calculated as,

$$T = F \times r$$

Where
   T = Torque in Newton meters
   F = Force applied by brake
   r = Radius from center of motor shaft to point of force application.

A voltmeter and an ammeter are connected for measuring the supply voltage and the current taken by the motor.

A load is applied gradually to the motor and the corresponding current, voltage, torque, and speed-reading are noted. In the case of testing a series motor, it is better to take the readings from a high load to gradually reducing load.

As the whole output is wasted in heat, it is used only for small capacity motors. A measurement of force, exerted by the arm and the radius or length of the arm, can give the torque.

If R is the radius of the pulley, and W and w are weights on either side of the brake arrangement, then

$$\text{Net torque} = (W - w)\,R \text{ kgm}$$
$$= 9.81 \times (W - w)\,R \text{ N.m.}$$

$$\text{Output} = 2\,\pi n\,T/60 \text{ W}$$

The input to the motor is measured. If V is the measured applied voltage and I the input current, then

$$\text{Input} = V \times I \text{ W}$$

The efficiency of the motor:

$$\text{Efficiency} = \frac{\text{Output}}{\text{Input}}$$

A dynamometer arrangement can be used alternatively to measure the pull and the torque on the motor under braking conditions.

2. *Swinburne test*

This test is for shunt and compound type of DC machines and gives the efficiency of the motor on any load, from the data calculated on 'no load' conditions.

This is an indirect method of testing the efficiency of a DC shunt motor, by measuring losses, like iron loss and mechanical losses. In addition, it is assumed that these losses remain constant during all load conditions.

The DC motor is run on 'no load'. The voltage and speed are adjusted to the rated values. A connection diagram for the Swimburne test is shown in Figure 4.25.

The armature resistance (R_a) is measured with an ammeter/voltmeter method, including all of the series and interpole winding. Similarly, the shunt field resistance (R_f) is also measured.

The motor is run at a rated speed and at a rated voltage without any load.

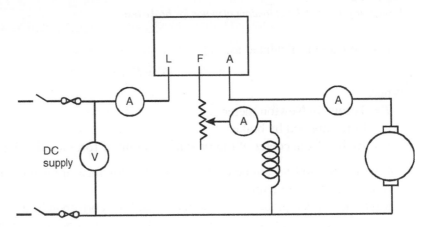

Figure 4.25
Test arrangement for Swinburne test

At no load:

Armature current = I_o
Input to the armature = $V I_o$
Since motor is at no load, this is the power required to overcome losses.

So, power loss

$$P_o = V \times I_o$$

Field copper losses is constant at all loads and is given by:

$$I_f^2 \times R_f$$

Constant losses at all loads

$$P_\text{o} + I_\text{f}^2 \times R_\text{f}$$

When motor is running armature losses is given by:

$$I_\text{a}^2 \times R_\text{a}$$

So total losses at any load, i.e., armature current is given by:

$$P_\text{L} = P_\text{o} + I_\text{f}^2 \times R_\text{f} + I_\text{a}^2 \times R_\text{a}$$

The efficiency of the DC motor is determined as:

$$\text{Efficiency} = \frac{(\text{Input} - \text{Losses})}{\text{Input}}$$

$$\text{Efficiency} = \frac{V \times (I_\text{a} + I_\text{f}) - P_\text{L}}{V \times (I_\text{a} + I_\text{f})}$$

Where P_L = Total losses at any armature current.
The following are the advantages of the Swinburne test:

- The efficiency of the machine can be determined without any direct loading. Energy is saved during the testing. This is quite useful, particularly for large machines.
- The efficiency can be determined at any load.

The following are the disadvantages of the Swinburne test:

- Only shunt and compound machines can be tested by this method. The series machines cannot be tested as they cannot run on no load.
- The effect of commmutation and armature reaction is not considered or tested in this method. The errors due to assumption of constant iron losses at all the loads are insignificant.

3. *Hopkinson test on shunt motors*

The direct loading method for large machines involves huge power losses. To avoid this, the regenerative method of testing is used. This test can be performed on two identical shunt machines, mechanically coupled to each other. One machine is run as the motor and the other as a generator. A typical arrangement for the Hopkinson test on two similar shunt machines is shown in Figure 4.26.

One machine starts as a motor taking supply from the mains. The field current is adjusted to run the machine at the rated speed. The second machine is mechanically coupled and is run at the same speed. The excitation of the machine is so adjusted, that the voltage across the armature is slightly higher than the supply voltage. This is checked with a Voltmeter V. The polarities of the machines should be suitable for a parallel operation. When the voltmeter reads zero or a slightly higher 1 or 2 V indicating that it would be generator action, the switch is closed so that both the machines are in parallel across the supply. As both the machines are similar and equal in size and rating, the constant losses – friction, windage, and iron – are assumed to be equal.

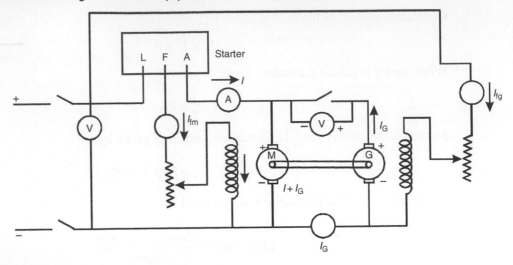

Figure 4.26
Test arrangement for the Hopkinson test on two similar shunt machines

The efficiency of the shunt machine acting as a motor and as a generator can be determined as follows:

For motor:

$$\text{Input} = V \times I_m + V \times I_{fm}$$

$$I_m = I_g + I$$

$$\text{Armature copper loss} = I_m^2 \times r_{am}$$

$$\text{Field copper loss} = V \times I_{fm}$$

$$P_m = \frac{(V \times I - I_g^2 \times r_{ag} - I_m^2 \times r_{am})}{2}$$

Where P_m = Constant loss.

Total losses = armature copper loss + Field copper loss + Constant loss
Output = Input – Losses
Efficiency = Output / Input.

4. *Retardation test*

If a motor is brought up to the speed and then switched off, it will slow down typically, as shown in Figure 4.27.

The angular retardation dowdy at any instant is directly proportional to the rotating torque and it varies inversely as the moment of inertia (J) of the motor. The power consumed in overcoming the losses due to rotation is given as follows:

$$P = \frac{d(\tfrac{1}{2} \times J \varpi^2)}{dt}$$

or

$$P = J \varpi \times \frac{d\varpi}{dt}$$

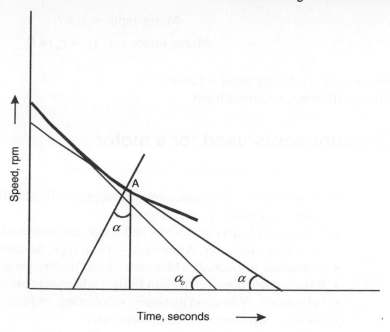

Figure 4.27
Retardation test on a motor

The test can be carried out on a separately excited motor or shunt motor. The friction losses and iron or core losses can be determined by this test.

5. *Field's series test*

This test is used for obtaining the efficiency of two similar motors. This test avoids the difficulty of obtaining readings on light loads, by using the motor current to excite the field as a generator. The generator armature is connected to the load resistance. The connection diagram for Field's test is shown in Figure 4.28. As both the machines have the same excitation, iron losses in both machines are considered equal.

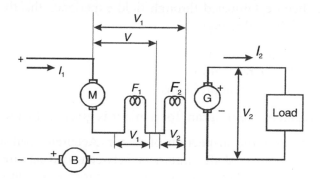

Figure 4.28
Field's efficiency test on two series motors

The total losses in each machine is given as:

$$P_0 = \frac{1}{2} \times \left[V \times I_1 - V_2 \times I_2 - I_1^2 (r_a + r_{se}) - I_2^2 \times r_a \right]$$

$$\text{Motor input} = V \times I_1$$
$$\text{Motor losses} = I_1^{2}{}'(r_a + r_{se}) + P_o$$

Motor output = Motor input − Losses
Motor efficiency = Output/Input

4.11   Measurements used for a motor

The following are the measurements used for a motor:

- *Temperature*:   Thermocouples/Resistance elements/thermistors measure temperatures of windings, bearings, etc.
- *Voltage and current*:   Volts and Amps are measured using portable voltmeter, recording voltmeter, Ammeters – clamp type, recording, CRO, etc.
- *Insulation resistance*:   Meggers – hand/motor-operated.
- *Winding resistance*:   Kelvin bridge, wheatstone bridge, resistance meters, etc.
- *Vibration*:   Vibration metering, monitoring, and analyzing equipment.
- *Speed*:   Stroboscopes, tachometer, etc.
- *Dielectric loss angle measurement*:   Tan–delta measurement.

4.12   Motor failures and methods to extend its life

The trouble free operation of induction motors over the greater part of their service life, which in many instances, exceeds that of driven equipment, requires little more than regular and routine maintenance chores. Regular cleaning, correct lubrication, and proper maintenance, is all that is required to ensure a consistently high level of performance from a motor, that is correctly selected and properly installed.

In essence, the useful service life of a motor is largely a function of the quality of maintenance. Maintenance is all the more important, in the context of the present-day motors as they are precisely designed to exact ratings and optimized parameters. Hence, any lapse in the proper maintenance of the motors is likely to affect the performance.

It has been established through field experience that the majority of the failures occur because of the following:

- Insulation failures
- Rotor-bar failures
- Mechanical problems.

The maintenance program for a motor is given as follows:

1. Periodic inspection of motor. Accurate shaft alignment. For directly coupled motors, shaft alignment between load and motor shaft should be proper. In case of belt-type system, check for belt condition, belt tension.
2. Check motor heating. If motor heats up quickly, check and clean air filters. Therefore, the airflow will be adequate.
3. Keep motor clean and free from dirt and oil.
4. Check for dampness around the motor or inside the motor. This can reduce insulation strength of motor winding. As far as possible, keep motor dry internally as well as externally. Also, run motor for few hours if not in use for a long time so that moisture dries.

5. Check bearing condition on a regular basis. Bearing should be lubricated with prescribed lubricant. At the same time, keep in mind that lubrication should be always done in proper quantity. Excess as well as lesser quantity can do harm.
6. Check for any abnormal noise or excess vibrations from motor or coupling. Do vibration analysis if necessary.

If the above guidelines are followed, the motor will remain problem free.

Motors driven by variable speed drives have specific requirements for correct performance such as special cooling, bearing insulation, use of terminal filters to absorb high voltage pulses, etc. In the case of retrofitted drives, the manufacturer needs to be consulted and any additional measures recommended need to be incorporated to avoid failures of motor.

4.13 Motor control trouble–remedy table

Trouble	Cause	Remedy
Contacts chattering	1. Poor contact in control circuit 2. Low voltage	1. Replace the device 2. Check coil terminal voltage/general voltage fluctuation/voltage dips during starting
Welding or freezing	1. Current inrush abnormal 2. Tip pressure low 3. Low voltage 4. Ingressed foreign matter preventing contact closing 5. Short-circuit/ground fault	1. Check shorts/grounds. Check motor load current. Use higher size contactor 2. Replace contacts/springs. Contact carrier may be damaged 3. Check coil terminal voltage and voltage dips during starting 4. Clean contacts with Freon 5. Remove fault. Ensure correctly rated fuse/circuit breaker is used
Short tip life tip overheating	1. Filing or dressing 2. High current interruption 3. Low tip pressure 4. Foreign matter ingress 5. Short-circuits/ground fault 6. Loose power circuit connection 7. Persistent overload	1. Ensure silver tips are not filed. Rough spots or discoloration do not harm tips or cause malfunction 2. Replace with higher size device. Check current levels/faults 3. Replace contacts/springs. Contact carrier may be damaged 4. Clean contacts with Freon. Check enclosure for ambient condition suitability 5. Remove fault. Check correctly rated fuse/circuit breaker is used 6. Tighten 7. Check motor load current. Install larger device

(Continued)

Coils open circuit	1. Mechanical damage	1. Handle/store coils with care
Overheated coil	1. Overvoltage or high ambient temperature 2. Coil unsuitable 3. Shorted turns due to mechanical damage 4. Undervoltage/magnet seals in failure 5. Pole faces dirty 6. Obstruction to moving elements	1. Check terminal voltage less than 110% of rated voltage 2. Replace with correct coil 3. Replace coil 4. Check coil terminal voltage. This should be at least 85% of rated voltage 5. Clean pole faces 6. Check free movement of contact and armature assembly
Overload relay Tripping	1. Persistent overload 2. Corrosion or loosening 3. Unsuitable thermal units 4. High coil voltage	1. Check excessive motor currents, current unbalance. Take corrective action 2. Clean/tighten 3. Replaced with correct size for the application and conditions 4. Check coil voltage is within 110% of rated capacity
Trip failure	1. Thermal units not suitable 2. Mechanical bindings, dirt, corrosion, etc. 3. Damaged relay 4. Relay contact welded	1. Apply proper thermal units 2. Clean/remove particles/obstruction, etc. to restore to proper functioning condition. Replace relay/thermal unit if not possible 3. Replace relay and thermal units 4. Replace contact or entire relay as necessary
Magnetic and mechanical parts noisy magnet	1. Shading coil broken 2. Magnet faces dirty/rusty 3. Low voltage	1. Replace magnet and armature assembly 2. Clean 3. Check coil terminal voltage/voltage fluctuation/motor starting voltage dips

| Pickup and sealing failure | 1. Loss of control voltage

2. Low voltage

3. Moving part obstruction

4. Open or overheated coil
5. Coil unsuitable | 1. Check control circuit. For loose connection/poor contact continuity
2. Check coil terminal voltage/voltage fluctuation/ motor starting voltage dips
3. With power off, check contact and armature assembly movement
4. Replace
5. Replace |

Trouble	Cause	Remedy
Failure to drop out	1. Sticky substance on pole face 2. Voltage persistence 3. Worn or corroded parts failing to separate	1. Clean 2. Check coil terminal voltage/control circuit 3. Replace defective parts
	4. Residual magnetism caused by lack of air gap in magnet path. 5. Welding of contacts.	4. Replace magnet and armature 5. See 'Contacts – Welding or freezing'
Pneumatic timers erratic timing	1. Ingress of foreign matter in valve	1. Replace complete timing head. Return timer to factory for repair and adjustment
Failure of contact operation	1. Actuating screw not correctly adjusted 2. Worn/broken parts in snap switch	1. Adjust as per manual service instructions 2. Replace switch
Limit switches damaged parts	1. Actuator overtravel	1. Use resilient actuator. Operate within device tolerance limits
Manual starters failure to reset	1. Latching mechanism damaged	1. Replace starter

4.14 Motor starter check chart

Trouble	Cause	Corrective Action
Contactor/Relay closing failure	Supply voltage failure Low voltage Open-circuited coil Pushbutton, interlocks, or relay contact not making Loose connections or broken wire Incorrect pushbutton connection Open o/l relay contact Mechanical parts damaged, corroded, not properly aligned/assembled, etc.	Check fuses/disconnect switch. Check power supply. Ensure correct size of wire Replace Adjust to ensure correct movement, easy operation, and correct contact pressure Check circuit. Isolate circuit first Check with wiring diagram Reset relay Clean/align and adjust for proper operation

(Continued)

Trouble	Cause	Corrective Action
Contactor or relay fails to open	Incorrectly connected pushbutton	Check connection with wiring diagram and rectify
	Worn shim in magnetic circuit	
	Residual magnetism holds armature closed	Replace shim
	Pushbutton, interlock, or relay contact fails to open coil circuit	Make adjustment for correct movement, ease of operation, and proper opening
	'Sneak' circuits	Check for insulation failure
	Welding of contacts	See 'Excessive corrosion of contacts'
	Mechanical part malfunction due to damage corrosion, etc.	Clean mechanical parts. Check for free movement
		Remove obstruction/ingressed matter. Repair or replace worn or damaged parts
Contact corrosion/welding	Contact spring pressure not adequate. Overheating or arcing on closing	Adjust for correct contact pressure. Replace spring or worn contacts if necessary
	Reduction of effective contact surface area due to pitting, etc.	Dress up contacts with fine file. Replace if badly worn
	Abnormal operating conditions	Check rating and load. In case of severe operating condition replace open contactors with oil-immersed or dust-tight equipment
		Instruct operator in proper control of manually operated device
	Chattering of contacts due to external vibrations	Check control switch contact pressure. Replace spring if it fails to give rated pressure
	Sluggish operation	Tighten all connections. If problem persists mount/ move control, so that vibrations are decreased
		Clean and adjust mechanically. Align bearings. Check free movement
Arc lingers across contacts	Blow out problem	Check blow out type with wiring diagram. Check blow out circuit
	Series blow out may be short-circuited	
	Shunt blow out may be open	
	Ineffective blowout coil	Check rating. Replace in case of improper application. Check polarity and reverse coil if necessary

Trouble	Cause	Corrective Action
	Note travel of contacts, in case blowout is not used	Increasing travel of contacts increases rupturing capacity
	Arc box might be left off or not in correct position if blow out is used	Ensure that arc box is fully in place
	Overload	Check rating against load
Noisy AC magnet	Improper assembly	Clean pole faces. Adjust mechanical parts
	Broken shading coil	Replace
	Low voltage	Check power supply. Check wire size
Frequent coil failure	High voltage	Check supply voltage against controller rating
	Gap in magnetic circuit.	Check travel of armature. Adjust magnetic circuit. Clean pole faces
	Ambient temperature may be high	Check controller rating against ambient temperature. Replace coil with correctly rated coil for ambient, from manufacturer
Burning of panel/equipment due to starting resistor heat	Frequent starting	Use higher-capacity resistor

5

Switches, circuit breakers and switchboards

Objectives

- To understand switch gear types
- To understand protection types
- To understand the basics of power distribution.

5.1   Introduction

To open or close an electrical circuit, a switch device is required. In addition, to protect an electrical circuit, a protection device is required.

Low-voltage switches with HRC ceramic enclosed fuses are used commonly in industries though the present trend is to use circuit breakers with overload and short-circuit protection in place of the fuse switch. There are various types of switches and protection devices used in the industry for different applications. All these devices are generally called Switchgear.

Apart from switching on or off any section of an electrical installation, the switchgear must include the necessary protective devices. These protective devices automatically isolate a particular section of the installation under fault conditions.

The switchgear must withstand short-circuit faults without thermal or mechanical damage and are therefore given a short time withstand rating (usually 1 s in LV), Though this rating is mainly applicable to the bus bars and other conductors, control devices such as switches and circuit breakers must also have such rating (either short time or dynamic withstand for prospective short-circuit current) appropriate to their function.

5.2   Switches and circuit breakers

5.2.1   Switches

Knife switches are used for low-voltage circuits. These are mounted in front of the board or panel, and are operated by hand. Knife switches should be mounted for a vertical throw, with the blade side of the switch either dead or disconnected from the source of the power when open. This is to minimize the risk of an accidental contact.

Originally, all switchgears consisted of open knife switches. Protective devices such as fuses were mounted close to the switch. The use of high-voltage AC and the great increase of the total power in a system necessitated the use of oil-break, air-break, vacuum, air blast, or SF6 switchgear.

In LV installations knife type switches are of metal enclosed or cubicle mounted, double break type complete with arc chutes. The off-circuit LV isolators have been largely replaced by switches of either load break or load-break fault-make capabilities. In some applications, open-type boards are installed, but generally, most of the switchgear today is enclosed. Knife switches are usually spring controlled, giving a quick make and break with a free handle action. This makes the operation of the switch independent of the speed at which the handle is moved.

In all cases, it is impossible to open the cover with the switch in the on position. The rated current capacities of LV cubicle type switches with independent manual operation are limited to 630 A with some vendors even offering switches of 800 or 1000 A on request.

Copper-brush switches substitute a leaved copper brush with a wiping contact for the knife-blade contact, and make use of an auxiliary break, between the carbon blocks, to prevent burning of the copper leaves due to arcing. This type of a switch has been used as a circuit breaker, particularly in MV range with remote action by the addition of tripping coils, though closing is not done remotely. Switches with integral MV fuses also have the provision to open the switch on fuse blowing.

5.2.2   Circuit breakers

A circuit breaker works as a switching device as well as a current interrupting device. It does this by performing the following two functions:

1. Switching operation during normal working of operation and maintenance
2. Switching operation during abnormal conditions that may arise, such as over-current, short-circuit, etc.

Therefore, the need arises that it must withstand the abnormal current conditions, apart from the normal working current. All the switches discussed above, come equipped with a tripping device that constitutes an elementary load interrupter switch. The difference between a load interrupter switch and a circuit breaker lies in the current interrupting capacity.

A circuit breaker must open the circuit successfully under short-circuit conditions. The current through the contacts may be several orders of magnitude greater than the rated current. As the circuit is opened, the device must withstand the accompanying mechanical forces and the heat of the ensuing arc, until the current is permanently reduced to zero.

When any high-voltage circuit is interrupted, there is a tendency towards an arc formation between the two separating contacts.

If the action takes place in air, the air is ionized and plasma is formed by the passage of current. When ionized, the air becomes an electric conductor. The space between the separating contacts thus has a relatively low impedance and the region close to the surface of the contacts has a relatively high voltage drop. The thermal input to the contact surface is therefore relatively large and can be highly destructive.

Therefore, the major aim in a circuit breaker design is to quench the arc rapidly enough, to keep the contacts in a reusable state by one of the following methods:

1. *High-resistance interruption*

In this method, arc resistance is increased. This method is generally used in DC circuit breakers and low-medium-voltage AC circuit breakers. The increase of arc resistance is caused by elongation of the arc against an arc chute which contains arc-splitting plates.

The arc is driven outwards using a combination of contact profile, air movement and in some cases by a magnetic blow out device.

2. *Low-resistance of zero-point extinction*

In this method, the arc is interrupted at a current zero instance. At that instance, the air between the separating contacts is unionized by introducing fresh air, SF6 gas, or oil between the contacts. Naturally, this method is used for an AC arc interruption.

By using a combination of shunt and series coils, the circuit breaker can be made to trip when the energy reverses. Circuit breakers may trip, when a local breaker or fuse immediately clears the difficulty.

To ensure that the service is uninterrupted, automatic re-closing schemes are often used for circuit breakers feeing to overhead lines where self clearing faults (e.g. a bird fault) can occur. After tripping, an automatic scheme operates to re-close the breaker with a short delay giving an opportunity for the fault to clear. If a short-circuit still exists, the breaker trips once again. The breaker attempts to re-close two to three times and if the short-circuit persists it remains permanently locked out.

5.2.3   Miniature circuit breakers

Miniature circuit breakers (MCBs) are used widely, as protective devices for switching and protection in domestic, commercial, and industrial applications. They are popular because they replace the conventional fuse-and-switch and give more flexibility.

During a normal operation, it works as a switch; while during overload or short-circuit condition, it works as a protection device, isolating the faulty section.

Magnetic or thermal sensing devices provided within it actuate a tripping mechanism.

Typical voltage ratings: 240 V/415 V AC; 50 V/110 V DC
Typical current rating: 1–55 A

5.2.4   Molded case circuit breakers

These are circuit breakers with tripping mechanisms and terminal contacts assembled together in a molded case.

This helps in getting high die-electric strength as well as mechanical strength to it. In addition, an arc chute is provided to increase the length of the arc and at the same time restricting hot gases that come in contact with the important parts of the breaker.

Molded case circuit breakers (MCCB) with ratings up to 3000 A are capable of interrupting currents up to 200 kA. These are used for the control of low-voltage networks.

5.2.5   Oil circuit breakers

The arc decomposes in dielectric oil. The gases formed due to the decomposition are channeled through a vent in the chamber.

Oil circuit breakers are popular for high-voltage distribution systems, despite the perceived fire risk. These consist of an oil enclosure, in which the contacts and an arc control device are mounted. The arc is kept within the control device and the resultant gas pressure sweeps the arc, through the cooling vents in the side of the pot. There is a possibility of an explosion due to rise in pressure. In addition, these circuit breakers require regular replacement of oil as dielectric strength reduces during arcing.

They are not suitable for applications where breakers operate repeatedly. The oil circuit breakers are commonly used up to the voltage level of 145 kV.

5.2.6   Air-circuit breakers

In this type of breakers, air (at atmospheric pressure) is used for arc extinguishing. It uses the high-resistance interruption principle. The length of an arc is increased by using arc chutes and arc runners.

They are used in AC and DC circuits up to 11 kV. They are generally of an indoor type for medium- and low-voltage applications. They are simple in construction, indoor type, panel-mounted and have current-limiting properties. They are particularly suitable for applications where repeated breaking is required. The operation can be manual as well as automatic.

The manual operating mechanisms can be either by spring or by motor charging, whereas in the automatic mode it can be through the solenoid coil.

The air-break circuit breaker for 3.3–11 kV has an arc control device, which is suitable for motor switching and is used mainly in power stations.

Typical characteristic curves of an air-circuit breaker are shown in Figure 5.1.

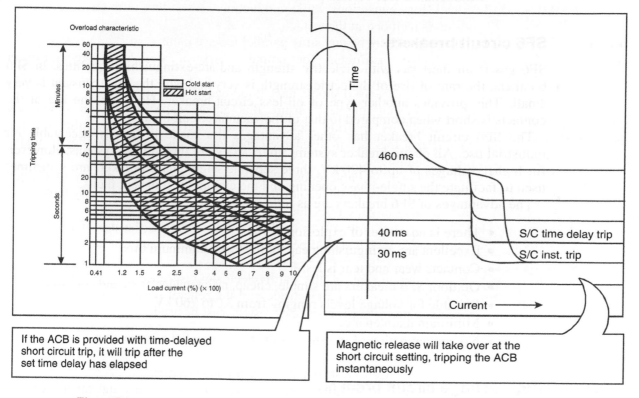

Figure 5.1
Typical characteristics curves of an air-circuit breaker

5.2.7   Vacuum circuit breakers

When two current-carrying contacts are separated in a vacuum chamber, an arc is drawn between them. With an AC, the current decreases during a portion of the wave and then tends to zero. This principle is used in these breakers.

Vacuum circuit breakers were the first type of oil-less circuit breakers. Vacuum interrupters are sealed for life in ceramic bottles. These bottles contain movable contacts in a high vacuum. The circuit breaking performance of this design is very high and a large number of short-circuit operations can be achieved before any replacement is necessary.

Vacuum circuit breakers are used for duties that require the following:

- Very high electrical and mechanical life
- Concealed current path
- Compactness.

The advantages of a vacuum circuit breaker are listed as follows:

- No exposed arc
- High operating safety due to reliable switching on high short-circuit faults
- Longer electrical life of up to 30 000 operating cycles at rated current
- High short time current withstanding capability
- Extremely short fault clearance time
- Integrated voltage suppressor
- Contact erosion indicator for erosion monitoring
- Maintenance-free current path.

5.2.8   SF6 circuit breakers

SF6 gas is an inert gas with dielectric strength and arc-extinguishing qualities. In SF6 breakers, the rate of rise of dielectric strength is very high and the time constant is very small. This provides another type of oil-less circuit breaker. However, the life of the contacts is short when compared to that of the vacuum circuit breaker.

The SF6 circuit breaker has other advantages that make it equally acceptable for industrial use. All circuit breaker systems up to 36 kV are three-phase systems. However, for higher voltages of up to 420 kV, three separate single-phase breakers are sometimes used to facilitate the single-phase opening and the closing for transient faults.

The advantages of SF6 breakers are as follows:

- There is no danger of explosion or fire
- Excellent arc-extinguishing capabilities with minimum time
- Contacts wear and tear is lesser
- Outdoor SF6 breakers are simple, cheap, maintenance-free, and compact
- Suitable for voltage levels ranging from 3.6 to 760 kV
- Minimum maintenance
- No contamination of moisture or dust due to sealed construction.

5.2.9   High-voltage circuit breakers

High-voltage circuit breakers are either of the oil type, in which the contacts open under oil, or of the air-blast type. In this type of breakers, high-pressure air is forced on the arc through a nozzle at the instant of contact separation. The portion of ionized air between the contacts is blown away by the blast of high-pressure air. Therefore, they are called either air-blast circuit breaker or compressed air-circuit breaker.

A CT, on an inverse-time relay in which the time of closing the relay contacts is an inverse-time function of the current, initiates tripping of the high-voltage circuit breaker. Therefore, the greater the current, the shorter is the time of closing.

When the DC circuit is closed by a relay contact, a DC tripping coil trips the breaker. The circuit breakers should open the circuit within 6 cycles from the time of closing of the relay contacts. Air-blast circuit breakers have received a wide acceptance in all fields for both indoor and outdoor applications. Indoor breakers are available up to 40 kV and interrupting capacities of up to 2.5 GVA. Outdoor three-pole breakers are available in extra-high-voltage ratings of up to 765 kV, capable of interrupting 55 GVA or 40 000 A of symmetrical current.

5.2.10   Motor circuit breakers

Motor circuit breakers provide overload, short-circuit, and single-phase protection for three-phase motors. The motor circuit breaker has a toggle switch for ease of operation and has auxiliary contacts, trip indicating contacts, and a U/V or shunt release.

The three-pole circuit breaker can be connected in parallel to the fuses. In the event of one fuse blowing, the breaker actuated through its release gives a tripping signal through its auxiliary contacts to the motor control device to switch off the motor. Thus, the motor is not subjected to single phasing, and costly motor burnouts are prevented.

The motor circuit breakers operate on the current-limiting principle. In the case of a short-circuit, the contacts are opened electro-dynamically by the short-circuit current. The instantaneous over-current releases, through the switching mechanism trips all the three poles of the breaker. A large arc voltage is quickly built up in the arc chamber, limiting the short-circuit. The breakers have a trip-free mechanism, and tripping cannot be prevented by the toggle switch position. After clearing the fault that caused the short-circuit, the limiter must be reset by hand before the circuit breaker can be switched on again.

Typical characteristic curves for overload and short-circuit release, and the current-limiting feature of a motor circuit breaker are shown in the Figure 5.2.

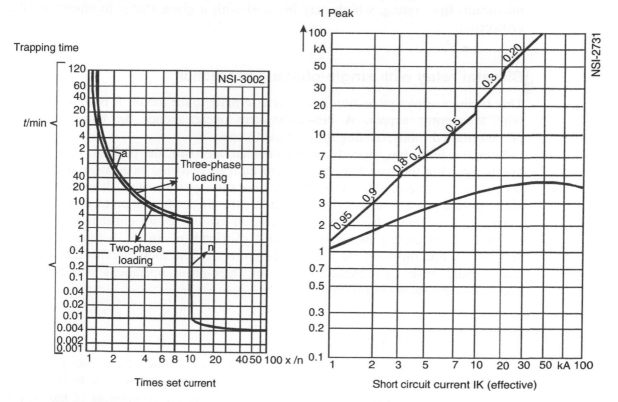

Figure 5.2

Typical characteristic curves of a motor circuit breaker for overload and short-circuit release and current-limiting feature

5.3 Overloads and fault protection

Protection devices against electrical faults may be broadly divided into fuses or circuit breakers. In some applications, fuses are used with the circuit breakers to take over the interruption of higher short-circuit currents, particularly with the miniature or lower-rated MCCB.

5.3.1 Overload and fault protection in motor circuits

Often a motor is loaded beyond its rated capacity due to incorrect operating conditions. This leads to a motor overload, an increase in current flowing through the winding and an increase in the temperature of the winding. This results in a permanent damage to the motor winding and the cables.

In a motor circuit, the starter overload relays, protect the motor, and the associated cables against overload and the fuses in the circuit provide the required degree of short-circuit protection. A short-circuit protection is required to protect motor conductors, overload relays, and motors from the short-circuit condition. It is achieved by using the non-time delay fuse, instantaneous trip breaker, or the inverse time-breaker.

Usually, manufacturers give recommendations regarding the fuse ratings required to cope with the motor starting surges and indicate the minimum cable sizes required to achieve a short-circuit protection. In a well-designed combination, the starter itself interrupts all the overloads up to the stalled rotor condition. The fuses should only operate in the event of an electrical fault. The starter manufacturers indicate the maximum fuse rating, which may be used with a given starter to ensure satisfactory protection.

5.3.2 Bimetal relay with single-phasing protection

This is an overload protection provided externally to the motor. It is connected in series with the motor supply. A bimetallic strip operates once the temperature exceeds predetermined limits, causing the contacts to open.

After the relay has tripped and the contacts are open, the problem should be solved before pressing the reset button. The bimetal relays provide an accurate overload and an accelerated single-phasing protection for the three-phase motors. It incorporates a dual slider principle for accelerated tripping under the single-phasing protection.

The bimetal relay also provides protection against severe unbalanced voltages. The bimetal relays protect themselves against overloads of up to 10 times the maximum setting. Beyond this limit, they have to be protected from short-circuits. It is mandatory to use backup fuses. *I–t* characteristics for three-phase operations and single-phasing conditions are shown in Figure 5.3.

5.3.3 Phase failure relays

This protection interrupts power in all phases upon failure of any one phase. Normal overload relays or fuses may not protect the motor from damage due to single phasing. Phase-failure relay senses the negative-sequence voltage component of the supply and offers protection against phase failure, unbalanced phases, phase reversal, and under-and-over voltage faults.

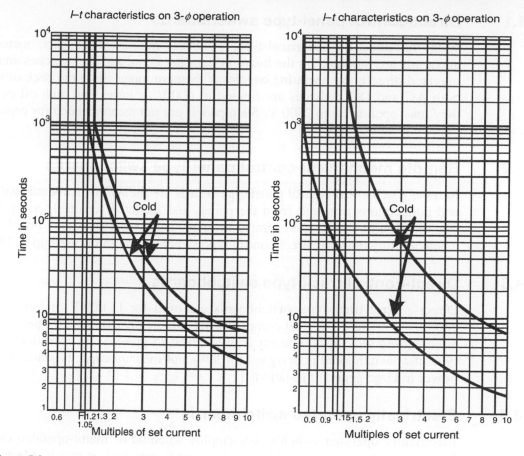

Figure 5.3
I–t *characteristics for three-phase operation and single-phasing condition*

5.3.4 Winding-protection relays

Winding-protection relays provide protection against overheating of the windings of motors, alternators, transformers, etc. Temperature is sensed with the help of a PTC thermistor embedded in the winding that gives a tripping signal when the temperature exceeds the response temperature of the thermistor.

In some cases, thermocouples or RTD (resistance temperature detectors) are fitted inside the winding to accurately indicate the temperature of the winding.

5.4 Switchboards

A switchboard is a distribution board (DB) that receives a large amount of power and dispatches it in small packets to various electrical equipments.

It has power-controlling devices such as breakers, switches along with protection devices such as fuses, etc.

Switchboards in general are divided into the following four classes:

- Direct-control panel-type
- Remote mechanical-control panel-type
- Direct-control truck-type
- Electrical-operated.

5.4.1   Direct-control panel-type switchboards

With the direct-control panel-type, switches, rheostats, bus bars, meters, and other apparatus are mounted on the board or near the board and the switches and rheostats are operated directly by operating handles if they are mounted on the back of the board. For both AC and DC, voltages are limited to 600 V or lower, but with oil circuit breakers, they may operate up to 2500 V. Such panels are not recommended for capacities of more than 3000 kVA.

5.4.2   Remote mechanical-control panel-type switchboards

Remote mechanical-control panel-type boards are the AC switchboards with the bus bars and connections removed from the panels and mounted separately away from the load. The oil circuit breakers are operated by levers and rods. This type of board is designed for a heavier duty than the direct-control type switchboards and is used up to 25 000 kVA.

5.4.3   Direct-control truck-type switchboards

Direct-control truck-type switchboards are used for 15 000 V or lower and consist of equipment enclosed in steel compartments completely assembled. The high-voltage parts are enclosed and the equipment is interlocked to prevent any operational mistakes. This type of a switchboard is designed for low- and medium-capacity plants and for auxiliary power in large generating stations.

5.4.4   Electrically operated switchboards

Electrically operated switchboards employ solenoid or motor-operated circuit breakers. Rheostats, etc. are controlled by small switches mounted on panels. Electrically operated switchboards make it possible to locate high-voltage and other equipment independent of the location of the switchboard.

Switchboards should be erected at least 1–2 m (3–4 ft) from the walls. Switchboard frames and structures should be grounded. For low-potential equipment, the conductors on the rear of the switchboard are usually made of a flat copper strip known as a copper bus bar. Aluminum bus bars are also used due to its low cost. Switchboards must be individually adapted for each specific electrical equipment/system.

5.5   Motor control center

In large plants, a number of electrical motors are placed offsite. To conveniently locate all supply cables, control circuitry, and various protections at one location, there is a Motor Control Center (MCC).

The size of an MCC depends on the number of electrical circuits and motors it controls. An MCC consists of a number of cubicles or compartments in a compact, floor-mounted assembly. The cubicles are sized differently for starters depending on the rating of the motor it controls.

6

Troubleshooting variable speed drives

Objectives

- To understand the basics of VSDs
- To understand the basics of converters
- To understand the basics of inverters
- To understand the installation, commissioning, and troubleshooting guidelines.

6.1 The need for VSDs

Fixed-speed motors and two-speed motors have been dealt with in previous chapters. Various industrial applications require motion control of machines with the help of such motors. VSDs provide continuous range control over the speed of the machines.

Some applications, such as paper mills, rolling mills, pumps, and machine tools cannot run without these while others, such as centrifugal pumps, can benefit from the energy savings. In general, VSDs are used to perform the following:

- Match the speed of a drive to the process speed requirements
- Match the torque of a drive to the process torque requirements
- Save energy and improve efficiency.

6.2 Basic VSD

Any basic electrical VSD consists of a motor, drive control unit, sensing unit, and an operator input.

The basic block diagram of a variable speed electrical drive is show in Figure 6.1.

The drive control unit is a device, which modulates power going from the source to the motor. Using the operator panel, one can increase or decrease the speed set point. A feedback unit gives the drive the actual speed feedback. The power modulator or the drive control unit then controls the speed, torque, and power, along with the direction of the motor and the machine. The power modulator may be used as a single device, for controlling a motor. It may have to be used in a combination type for certain other types of applications. The following are the types of power modulators or converters along with a brief description of each.

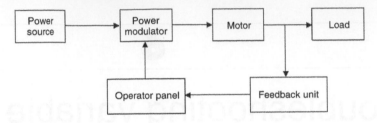

Figure 6.1
Block diagram of VSD

6.2.1   Converters

These convert one form of energy into another form, suitable for a motor. They can be defined as assemblies of power electronic components, which convert one or more of the characteristics of an electric power system.

For controlling DC motors, a variable DC voltage is required. For AC motors, a fixed frequency, variable AC voltage, or a variable frequency, variable voltage is required. To fulfill the requirement, the following devices are used. Given below are the various types of converters and their combinations:

1. *AC to DC converters*

These are classified as:

- Uncontrolled diode rectifier
- Half controlled rectifier
- Full controlled rectifier
- Rectifier with self-commutated devices.

In an uncontrolled rectifier, the fixed DC voltage at output is different from the AC supply at input. In a half-controlled rectifier, the variable DC voltage at output with a positive voltage and current is called a single quadrant drive. With a full controlled rectifier, the DC voltage of positive/negative polarity and the current in a positive direction is called a double quadrant drive. A full controlled rectifier has commutating devices such as the GTO (gate turn-off thyristors) and power transistors. It can be a single or a double quadrant drive.

If used with a full controlled rectifier, it can then provide four quadrant functionality, i.e., voltage as well as current in both directions.

2. *DC to DC converters*

Also known as choppers, these enable the variable DC voltage at output, from the fixed DC voltage given at input.

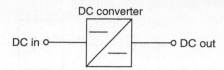

Choppers use devices such as the GTOs, thyristors, power MOSFETs, and IGBTs (insulated gate bipolar transistor).

3. *DC to AC converters or inverters*

The usage of inverters ensures a variable-frequency AC voltage at output from a fixed DC voltage given at input.

It is either a voltage source or a current source type. The output voltage or current can be changed along with the frequency by varying the DC input voltage. This occurs by feeding the DC voltage to the inverter through a rectifier. Variable voltage, variable-frequency AC voltage can be obtained using the PWM (pulse width modulation) for controlling the inverter.

4. *Cycloconverter*

The usage of cycloconverters ensures a variable voltage, variable-frequency AC voltage at output from a fixed voltage, and frequency AC voltage given at input.

These are built using thyristors, the firing of which is controlled by a control unit.

The following section details the various power electronic basic components. These components form a basic part of all circuits.

6.3   Power electronic components

Power electronic components are semiconductor devices, such as diodes, thyristors, transistors, etc. that are used in the power circuit of a converter. In power electronics, they are used in the non-linear switching mode (on/off mode) and not as linear amplifiers. In other words, these devices behave like an electronic switch.

An electronic switch electronically connects or disconnects an AC or DC circuit and can usually be switched ON and/or OFF. Conduction is usually permitted in one direction only.

Electronic switch

AC or DC in o———————|>|———————o AC or DC out

The following components are devices that are commonly used as electronic switches in power electronic converters. Developments in semiconductor technology have made these power electronic components smaller, more reliable, more efficient (lower losses), cheaper, and able to operate at much higher voltages, currents, and frequencies. The idealized operating principles of these components can be described in terms of simple mathematical expressions.

- Power diodes
- Power thyristors
- GTO
- MOS-controlled thyristors (MCT)
- Power bipolar junction transistors (BJT)
- Field effect transistors (FET, MOSFET)
- Insulated gate bipolar transistor (IGBT)
- Resistors (provide resistance)
- Reactors or chokes (provide inductance)
- Capacitors (provide capacitance).

6.3.1　Power diodes

A power diode is a semiconductor power on/off switch that allows flow of current in one direction, depending upon its connection. It is a two-terminal semiconductor device.

The two terminals of a diode are called the anode (A) and the cathode (K). These names are derived from the days when valves were commonly used.

Construction-wise, it has a single P–N junction. It consists of a two-layer silicon wafer attached to a substantial copper base. The base acts as a heat sink, a support for the enclosure and one of the electrical terminals of the diode. The other surface of the wafer is connected to the other electrical terminal. The enclosure seals the silicon wafer from the atmosphere and provides adequate insulation between the two terminals of the diode.

Symbol

Ideal

- *Forward conduction*:　Resistance less
- *Reverse blocking*:　Loss less
- *Switch on/off time*:　Instantaneous.

It is forward-biased, when the anode is positive, relative to the cathode and the diode conducts current, i.e., the switch is closed. It is reverse-biased, when the anode is negative, relative to the cathode and the flow of the current is blocked, i.e., the switch is open. This ability of the diode, to block the current flow in one direction, makes it suitable for rectifier applications, where it is required to allow the current flow in one direction only.

Depending on the application requirements, the following types of diodes are available:

Schottky diodes These diodes are used where a low forward voltage drop, typically 0.4 V, is needed for low output voltage circuits. These diodes have a limited blocking voltage capability of 50–100 V.

Fast recovery diodes These diodes are designed for use in circuits where fast recovery times are required, for example, in combination with controllable switches in high-frequency circuits. Such diodes have a recovery time (t_{RR}) of less than a few microseconds.

Line-frequency diodes The on-state voltage of these diodes is designed to be as low as possible to ensure that they switch on quickly in rectifier bridge applications. Unfortunately, the recovery time (t_{RR}) is long, but this is acceptable for line-frequency rectifier applications. These diodes are available with blocking voltage ratings of several kV and current ratings of several hundred kA. In addition, they can be connected in series or parallel to satisfy high-voltage or current requirements.

6.3.2 Power thyristors

Thyristors are sometimes referred to as SCR (silicon-controlled rectifiers). This was the name originally given to the device when it was invented by General Electric (USA) in around 1957. However, this name has never been universally accepted and used.

The name thyristor, is a generic term, that is applied to a family of semiconductor devices that have regenerative switching characteristics. There are many devices in the Thyristor family including the power thyristor, the GTO, the field controlled thyristor (FCT), the Triac, etc. It has two power terminals, called the anode (A) and the cathode (K), similar to a diode, and a third control terminal called the Gate (G), which is used to control the firing of the thyristor.

It is operationally similar to a diode, except that it requires a momentary positive voltage pulse, at the gate terminal, for conduction when connected in forward bias. A thyristor consists of a four-layer silicon wafer with three P–N junctions. High-voltage, high-power thyristors sometimes also have a fourth terminal, called an auxiliary cathode. This is used for connection to the triggering circuit. This prevents the main circuit from interfering with the gate circuit.

A thyristor is very similar to a power diode in both physical appearance and construction, except that the gate terminal is required to trigger the thyristor into a conduction mode.

Symbol

Ideal

- *Forward conduction*:   Resistance less
- *Forward blocking*:   Loss less (no leakage current)
- *Reverse blocking*:   Loss less (no leakage current)
- *Switch on/off time*:   Instantaneous.

The thyristor is turned off when it becomes reverse-biased and/or the forward current falls below the holding current. This must be controlled externally in the power circuit. Most SCRs have a heat sink for dissipating the heat generated during the operation.

Triacs

This is a different device from the thyristors category. Construction-wise, two SCRs are connected anti-parallel with each other. SCR conducts in a forward direction only, but Triac conducts in both directions. Therefore, if the output of a diode is a DC current when connected in an AC circuit, the output of a Triac is an AC current instead of a DC current. Triac has three terminals named MT1, MT2, and Gate. Triac can conduct in any direction with the gate pulse, either positive or negative. Triac can be used to vary the average AC voltage going to a load by changing the firing angle.

6.3.3   Gate-controlled power electronic devices

A number of gate-controlled devices have become available in the past decade. These are suitable for use as bi-stable switches on power inverters for AC VSDs. These can be divided into the following two main groups of components:

- Those based on Thyristor technology such as GTO and FCT
- Those based on Transistor technology such as the BJT, FET, and the insulated gate bipolar transistor (IGBT).

Gate turn-off thyristor (GTO)

A GTO thyristor is another member of the thyristor family. It is very similar in appearance and performance to a normal thyristor, with an important additional feature being that it can be turned off by applying a negative current pulse to the gate. GTO thyristors have high current and voltage capability and are commonly used for larger converters. This is especially true when self-commutation is required.

Symbol

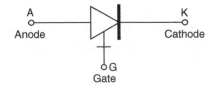

Ideal

- *Forward conduction*:   Resistance less
- *Forward blocking*:   Loss less (no leakage current)
- *Reverse blocking*:   Loss less (no leakage current)
- *Switch on/off time*:   Instantaneous.

The performance of a GTO is similar to a normal thyristor. Forward conduction is blocked until a positive pulse is applied to the gate terminal. When the GTO has been turned on, it behaves like a thyristor and continues to conduct even after the gate pulse is removed, if the current is higher than the holding current. The GTO has a higher forward voltage drop of typically 3–5 V. The latching and the holding currents are also slightly higher.

The important difference is that the GTO may be turned off by a negative current pulse applied to the gate terminal. This important feature permits the GTO to be used in self-commutated inverter circuits. The magnitude of the off pulse is large and depends on the magnitude of the current in the power circuit. Typically, the gate current must be 20% of the anode current. Consequently, the triggering circuit must be quite large and this results in additional commutation losses. Like a thyristor, conduction is blocked in the reverse-biased direction or if the holding current falls below a certain level.

Since the GTO is a special type of thyristor, most of the other characteristics of a thyristor covered above also apply to the GTO. The mechanical construction of a GTO is very similar to a normal thyristor with stud types common for smaller units and disk types common for larger units. GTO thyristors are usually used for high-voltage and current applications and are more robust and tolerant to over-current and over-voltages than power transistors. GTOs are available for ratings of up to 2500 A and 4500 V. The main disadvantages are the high gate current required to turn the GTO off and the high forward volt drop.

Field controlled thyristors (FCT)

Although the GTO is likely to maintain its dominance for high-power, self-commutated converter applications for some time, new types of thyristors are under development in which the gate is voltage-controlled. The turn-on is controlled by applying a positive voltage signal to the gate and the turn-off by a negative voltage. Such a device is called a FCT. The name affects the similarity to the FET. The FCT is expected to eventually supersede the GTO because it has a simple control circuit in which both the cost and the losses can be substantially reduced.

Power bipolar junction transistors (BJT)

Transistors have traditionally been used as amplification devices, where control of the base current is used to make the transistor conductive to a greater or a lesser degree. Until recently, they were not widely used for power electronic applications.

The main reasons were because the control and protective circuits were considerably more complicated and expensive, and transistors were unavailable for high-power applications. They also lacked the overload capacity of a thyristor and to protect transistors with fuses was unfeasible.

The NPN transistor, known as a BJT, is a cost-effective device for use in power electronic converters. Modern BJTs are usually supplied in an encapsulated module and each BJT has two power terminals, called the collector (C) and an emitter (E), and a third control terminal called the base (B).

Symbol

Ideal

- *Forward conduction*:   Resistance less
- *Forward blocking*:   Loss less (no leakage current)
- *Reverse blocking*:   Loss less (no leakage current)
- *Switch on/off time*:   Instantaneous.

A transistor is not inherently a bi-stable (on/off) device. To make a transistor suitable for conditions in a power electronic circuit where it is required to switch from the blocking state (high voltage, low current) to the conducting state (low voltage, high current) it must be used in its extreme conditions – fully off to fully on. This potentially stresses the transistor and the trigger, and protective circuits must be coordinated, to ensure the transistor is not permitted to operate outside its safe operating area. Forward conduction is blocked until a positive current is applied to the gate terminal and it conducts as long as a voltage is applied. During forward conduction, it also exhibits a forward voltage drop, which causes losses in the power circuit. The BJT may be turned off by applying a negative current to the gate.

Suitable control and protective circuits have been developed to protect the transistor against over-current when it is turned on and against over-voltage when it is turned off (Figure 6.2). When turned on, the control circuit must ensure that the transistor does not come out of saturation, otherwise; it will be required to dissipate a high power. In practice, the control system has proved to be cost effective, efficient, and reliable in service.

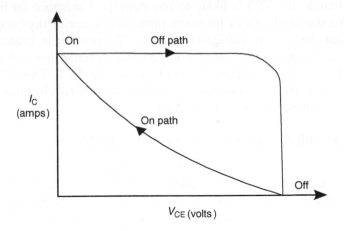

Figure 6.2
Desirable V–I *limits while switching a BJT*

The following are the advantages of BJT as a switch:

- Requires very low driving voltages
- Can operate at very high speed
- Can be turned on and off from the base terminal, which makes them suitable for self-commutated inverter circuits
- Good power handling capabilities
- Low forward conduction voltage drop.

The following are the disadvantages of BJT as a switch:

- Considered less robust and less tolerant of overloads and 'spikes' than thyristors
- Do not tolerate reverse voltages

- Relatively slow switching times compared with other devices
- Inferior safe operating area
- Has complex current-controlled gate driver requirements.

GTO thyristors are often preferred for converters. When BJTs are used in inverter bridges, they must be protected against high reverse voltages, by means of a reverse diode in series or in parallel. For the same reason, transistors are not used in rectifier bridges that have to be able to withstand reverse voltages.

The base amplification factor of a transistor is fairly low (usually 5–10 times). The trigger circuit of the transistor, consequently, must be driven by an auxiliary transistor to reduce the magnitude of the base trigger current required from the control circuit. To perform this, the Darlington connection is used.

Figure 6.3 shows a double Darlington connection, but for high-power applications, two auxiliary transistors (triple Darlington) may be used in cascade to achieve the required amplification factor. The overall amplification factor is approximately the product of the amplification factors of the two (or three) transistors.

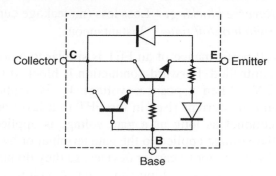

Figure 6.3
Power Darlington transistor

Transistors, used in VSD applications, are usually manufactured as an integrated circuit and encapsulated into a three-terminal module, complete with the other necessary components, such as the resistors and an anti-parallel protection diode. The module has an insulated base suitable for direct mounting onto the heat sink. This type of module is sometimes called a Power Darlington transistor module.

As shown in Figure 6.3, the anti-parallel diode protects the transistors from reverse biasing. In practice, this diode in the integrated construction is slow and may not be fast enough for inverter applications. Consequently, converter manufacturers sometimes use an external fast diode to protect the transistors. Power BJT are available for ratings of up to a maximum of about 300 A and 1400 V. For VSDs that require a higher power rating, GTOs are usually used in the inverter circuit.

Field effect transistor (FET)

BJT is a current-driven device. The current flows through the base controls and the flow of current is between the collector and the emitter. The FET Gate is voltage-controlled. FET is a special type of transistor that is particularly suitable for high-speed switching applications.

Its main advantage is that its Gate is voltage-controlled rather than current-controlled. It behaves like a voltage-controlled resistance with the capacity for high-frequency performance.

FETs are available in a special construction known as the MOSFET. MOS stands for metal oxide silicon. The MOSFET is a three-terminal device with terminals called the source (S), the drain (D), and the gate (G), corresponding to the emitter, collector, and gate of the NPN transistor.

Symbol

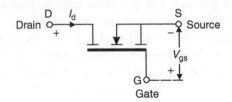

Ideal

- *Forward conduction*: Resistance less
- *Forward blocking*: Loss less (no leakage current)
- *Reverse blocking*: Loss less (no leakage current)
- *Switch on/off time*: Instantaneous.

The overall performance of an FET is similar to a power transistor, except that the gate is voltage-controlled. Forward conduction is blocked if the gate voltage is low, typically less than 2 V. When a positive voltage V_{gs} is applied to the gate terminal, the FET conducts and the current rises in the FET to a level dependent on the gate voltage. The FET will conduct as long as a gate voltage is applied. The FET can be turned off by removing the voltage applied to the gate terminal or by making it negative.

MOSFETs are majority carrier devices, so they do not suffer from long switching times. With their very short switching times, the switching losses are low. Consequently, they are best suited to high-frequency switching applications. A typical performance characteristic of a FET is shown in Figure 6.4.

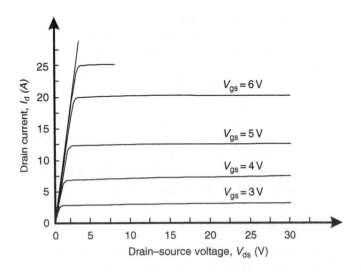

Figure 6.4
Typical characteristic of a FET

Initially, high-speed switching was not an important requirement for AC converter applications. With the development of PWM inverters, high-frequency switching has become a desirable feature to provide a smooth output current waveform. Consequently, power FETs were not widely used until recently.

At present, FETs are only used for small PWM frequency converters. The ratings are available from about 100 A at 50 V to 5 A at 1000 V, but for VSD applications, MOSFETs need to be in the 300–600 V range. The advantages and disadvantages of MOSFETs are almost exactly the opposite of BJTs.

The main advantages of a power MOSFET are given below:

- High speed switching capability (10–100 ns)
- Relatively simple protection circuits
- Relatively simple voltage-controlled gate driver with low gate current.

The main disadvantages of a power MOSFET are given below:

- Relatively low power handling capabilities
- Relatively high forward voltage drop, which results in higher losses than GTOs and BJTs, limits the use of MOSFETs for higher power applications.

Insulated gate bipolar transistor (IGBT)

The insulated gate bipolar transistor (IGBT) is an attempt to unite the best features of the BJT and the MOSFET technologies.

The construction of the IGBT is similar to a MOSFET with an additional layer to provide conductivity modulation, which is the reason for the low-conduction voltage of the power BJT.

The IGBT device has a good forward blocking but a very limited reverse blocking ability. It can operate at higher current densities than either a BJT or MOSFET by allowing a smaller chip size.

The IGBT is a three-terminal device. The power terminals are called the emitter (E) and the collector (C), using the BJT terminology, while the control terminal is called the gate (G), using the MOSFET terminology.

Symbol

Ideal

- *Forward conduction*:   Resistance less
- *Forward blocking*:   Loss less (no leakage current)
- *Reverse blocking*:   Loss less (no leakage current)
- *Switch on/off time*:   Instantaneous.

The electrical equivalent circuit of the IGBT shows that the IGBT can be considered as a hybrid device, similar to a Darlington transistor configuration, with a MOSFET driver, and a power bipolar PNP transistor. Although the circuit symbol above suggests that the device is related to a NPN transistor, this should not be taken too literally.

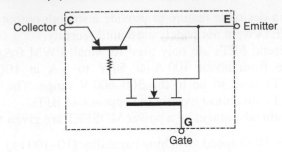

The gate input characteristics and gate drive requirements are very similar to those of a power MOSFET. The threshold voltage is typically 4 V. Turn-on requires 10–15 V and takes about 1 μs. The turn-off takes about 2 μs and can be obtained by applying 0 V to the gate terminal. The turn-off time can be accelerated, when necessary, by using a negative drive voltage. IGBT devices can be produced with faster switching times at the expense of an increased forward voltage drop.

IGBTs are currently available in ratings from a few amps to around 500 A at 1500 V, which are suitable for three-phase AC VSDs rated up to about 500 kW at 380 V/ 415 V/480 V. They can be used at switching frequencies up to 100 kHz. BJTs have now largely been replaced by IGBTs for AC VSDs.

The following are the main advantages of the insulated gate bipolar transistor (IGBT):

- Good power-handling capabilities
- Low forward conduction voltage drops of 2–3 V, which is higher than for BJT but lower than for a MOSFET of similar rating
- This voltage increases with temperature making the device easy to operate in parallel without danger of thermal instability
- High-speed switching capability
- Relatively simple voltage-controlled gate driver
- Low gate current.

Some other important features of the IGBT are given below:

- There is no secondary breakdown with the IGBT, giving a good safe operating area and low switching losses
- Only small snubbers are required
- The inter-electrode capacitances are not as relatively important as in a MOSFET, thus reducing Miller feedback.

There is no body diode in the IGBT, as with the MOSFET, and a separate diode must be added in anti-parallel when a reverse conduction is required, for example, in voltage source inverters.

6.4   Electrical VSDs

DC motors dominate in VSD applications due to their reliability. They also help to create a cheaper converter and control circuit.

As we have seen earlier, AC induction motors are mainly constant-speed motors. Since the 1980s, the popularity of the AC VSDs has grown rapidly, mainly due to advances in power electronics and digital control technology, affecting both the cost and the performance of this type of VSD. The main attraction of the AC VSDs is the rugged reliability and the low cost of the squirrel-cage AC induction motor compared to the DC motor.

As shown in Figure 6.5, changes in drives employed with respect to time have been shown. Figure 6.5 sections (a), (b), (c), and (d) are as per the following:

    (a) Ward-Leonard system
    (b) Thyristor-controlled DC drive
    (c) Voltage source inverter (PAM) AC drive
    (d) PWM voltage source (PWM) AC drive.

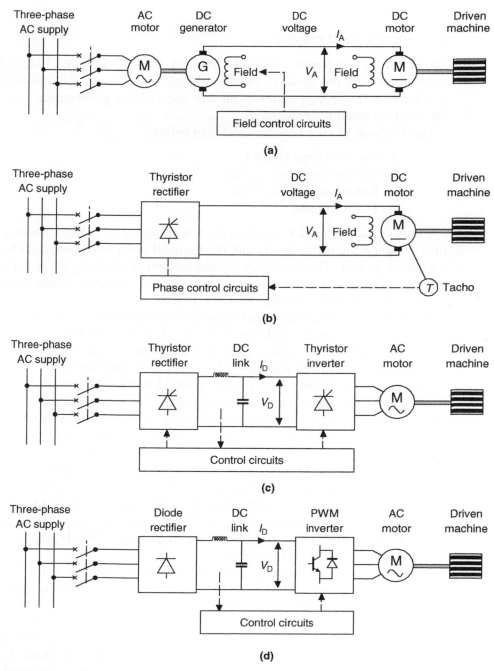

Figure 6.5
Main components of various types of VSD

The development path from the Ward-Leonard system to the thyristor-controlled DC drive and then to the PWM-type AC variable voltage, variable frequency converter is illustrated in Figure 6.5. In the first step, from (a) to (b), the high-cost motor-generator set has been replaced with a phase-controlled thyristor rectifier. In the second step, from (b) to (d), the high-cost DC motor has been replaced with a power electronic PWM inverter and a simple rugged AC induction motor. In an AC VSD, the mechanical commutation system of the DC motor has been replaced by a power electronic circuit called the inverter. However, the main difficulty with the AC VSD has always been the complexity, cost, and reliability of the AC frequency inverter circuit.

6.4.1   AC motor speed control

Developments in power electronics over the last 10–15 years has made it possible to control not only the speed of the AC induction motors but also the torque. Modern AC-VSDs, with flux-vector control, can now meet all the performance requirements of even the most demanding applications.

The methods of speed control are listed below:

1. Stator voltage control
2. Supply frequency control
3. Rotor resistance control
4. Pole changing.

Usually, the AC motor speed control is achieved by varying its supply frequency. In order to keep the losses minimum, the terminal voltage frequency is changed to keep the v/f ratio constant. The frequency control method of changing the speed of AC motors is a well-known technique for decades. Only recently, however, it has become a technically viable and economical method of VSD control.

AC drives have become a more cost-effective method of speed control, in comparison to DC drives, for most VSD applications of up to 1000 kW. It is also the technically preferred solution, for many industrial environments, where the reliability and the low maintenance, associated with the AC squirrel-cage induction motor are important.

The mains AC supply voltage is converted into a DC voltage and current through a rectifier. The DC voltage and current are filtered to smooth out the peaks before being fed into an inverter, where they are converted into a variable AC voltage and frequency. The output voltage is controlled, so that the ratio between the voltage and frequency remains constant in order to avoid over-fluxing the motor. The AC motor is able to provide its rated torque over the speed range of up to 50 Hz, without a significant increase in losses.

The motor can be run at speeds above the rated frequency, but with a reduced output torque. The torque is reduced because of the reduction in the air-gap flux, which depends on the *V/f* ratio. At frequencies below 50 Hz, a constant torque output from the motor is possible. At frequencies above the base frequency of 50 Hz, the torque is reduced in proportion to the reduction in speed.

One of the main advantages of VVVF (variable voltage variable frequency) speed control system is that, while the controls are necessarily complex, the motors themselves can be of a squirrel-cage construction, which is probably the most robust, and maintenance-free form of electric motor yet devised. This is particularly useful where the motors are mounted in hazardous locations, or in inaccessible positions, making routine cleaning and maintenance difficult. In locations that require machines to have flameproof or even waterproof enclosures, a squirrel-cage AC induction motor would be cheaper than a DC motor.

On the other hand, an additional problem with the standard AC squirrel-cage motors when used for variable speed applications is that they are cooled by means of a shaft-mounted fan. At low speeds, cooling is reduced, which affects the load ability of the drive. The continuous output torque of the drive must be de-rated for lower speeds, unless a separately powered auxiliary fan is used to cool the motor. This is similar to the cooling requirements of DC motors, which require a separately powered auxiliary cooling fan.

6.4.2   DC motor speed control

DC drives are widely used in the industry for their variable speed, good speed regulation, braking, and reversing ability.

In the past, DC motors were used in most of the VSD applications in spite of the complexity, high cost, and high maintenance requirements of the DC motors.

Even today, DC drives are still often used for the more demanding VSD applications. Examples of this are the sectional drives for paper machines, which require a fast dynamic response and a separate control of speed and torque.

Methods of speed control are given below

1. Armature voltage control
2. Field flux control.

Most of the DC drives employ armature voltage control method and field flux control, in order to achieve speed regulation, both below the rated speed and above the rated speed respectively. In both cases, the half- and fully controlled rectifier or converter is used to achieve a variable DC voltage, from AC voltage, to supply to the armature voltage.

Both the AC and DC drives use a converter or rectifier and inverter usually. The following section details such devices.

6.5   Power electronic rectifiers (AC/DC converters)

These devices convert a single or a three-phase AC power supply to a smooth DC voltage and current. Simple bi-stable devices, such as the diode and thyristor, can effectively be used for this purpose.

6.5.1   Assumptions

While analyzing power electronic circuits, it is assumed that bi-stable semiconductor devices, such as diodes and thyristors, are the ideal switches, with no losses and minimal forward voltage drop. It will also be assumed that the reactors, capacitors, resistors, and other components of the circuits have ideal linear characteristics with no losses.

Once the operation of a circuit is understood, the imperfections associated with the practical components can be introduced to modify the performance of the power electronic circuit. In power electronics, the operation of any converter is dependent on the switches being turned ON and OFF in a sequence. The current passes through a switch when it is ON and is blocked when it is OFF.

Commutation is the transfer of current from one switch turning OFF, to another turning ON. In a diode rectifier circuit, a diode turns ON and then starts to conduct current when there is a forward voltage across it, i.e., the forward voltage across it becomes positive. This process usually results in the forward voltage across another diode becoming negative, which then turns off and stops conducting current.

In a thyristor rectifier circuit, the switches additionally need a gate signal to turn them on and off. The factors affecting the commutation are illustrated in the idealized diode circuit in Figure 6.6, which shows two circuit branches, each with its own variable DC voltage source and circuit inductance. Assume, initially, that a current I is flowing through the circuit and that the magnitude of the voltage V_1 is larger than V_2. Since $V_1 > V_2$, diode D_1 has a positive forward voltage across it and it conducts a current I_1 through its circuit inductance L1. Diode D_2 has a negative forward voltage that blocks and carries no current.

Consequently, at time t_1

$$I_1 = I\sqrt{2}$$
$$I_2 = 0$$

Suppose the voltage V_2 is increased to a value larger than V_1, the forward voltage across diode D_2 becomes positive, and it starts to turn on. However, the circuit inductance L1 prevents the current I_1 from changing instantaneously and diode D_1 will not immediately turn off. Therefore, both the diodes D_1 and D_2 remain ON for an overlap period called the commutation time T_c.

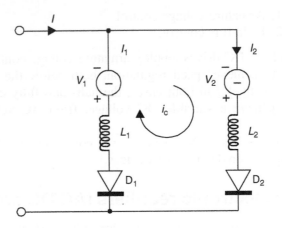

Figure 6.6
Simple circuit to illustrate commutation from diode D_1 to D_2

When both the diodes are turned on, a closed circuit is established which involves both branches. The effective circuit voltage $V_c = (V_2 - V_1)$, called the commutation voltage, drives a circulating current I_c, called the commutation current, through the two branches which have a total circuit inductance of $L_c = (L_1 + L_2)$.

In this idealized circuit, the voltage drop across the diodes and the circuit resistance has been ignored. From the basic electrical theory of inductive circuits, the current I_c increases with time at a rate dependant on the circuit inductance. The magnitude of the commutation current may be calculated from the following equations:

$$(V_2 - V_1) = (L_1 + L_2)\frac{di_c}{dt}$$

$$V_c = L_c \frac{di_c}{dt}$$

$$\frac{di_c}{dt} = \frac{V_c}{L_c}$$

If the commutation starts at a time t_1 and finishes at a time t_2, the magnitude of the commutation current I_c at any time t, during the commutation period, may be calculated, by integrating the above equation from time t_1 to t.

$$I_c = \frac{1}{L_c} \int V_c \, dt$$

During the commutation period, it is assumed that the overall current through the circuit remains constant.

$$I = (I_1 + I_2) \quad \text{constant}$$

As the circulating commutation current increases, the following hold true:

- Current (I_2) through the diode that is turning on increases in value

$$I_2 = I_c \quad \text{increasing}$$

- Current (I_1) through the diode that is turning off decreases in value.

$$I_1 = I - I_c \quad \text{decreasing}$$

For this special example, it can be assumed that the commutation voltage V_c is constant during the short period of the commutation. At time t, the integration yields the following value of I_c, which increases linearly with time.

$$I_c = \frac{V_c}{L_c}(t - t_1)$$

When I_c has increased to a value equal to the load current I at a time t_2, then the current has been transferred from branch 1 to branch 2, and the current through the switch that is turning off has decreased to zero. The commutation is then complete.
Consequently, at time t_2

$$I_1 = 0$$
$$I_2 = I_c = I$$

At the end of the commutation when $t = t_2$, putting I_c equal to I in the above equation, the time taken to transfer the current from one circuit branch to the other (commutation time), may be calculated as follows:

$$I = \frac{V_c(t_2 - t_1)}{L_c}$$

$$I = \frac{V_c t_c}{L_c}$$

$$t_c = \frac{I L_c}{V_c}$$

It is clear from the equation that the commutation time t_c depends on the overall circuit inductance ($L_1 + L_2$) and the commutation voltage. From this we can conclude the following:

- A large circuit inductance will result in a long commutation time.
- A large commutation voltage will result in a short commutation time.

In practice, a number of deviations from this idealized situation occur. The diodes are not ideal and do not turn off immediately when the forward voltage becomes negative. When a conducting diode is presented with a reverse voltage, some reverse current can still flow for a few microseconds, as indicated in Figure 6.7. The current I_1 continues to decrease beyond zero to a negative value before returning to zero. This is due to the free charges that must be removed from the PN junction before blocking is achieved.

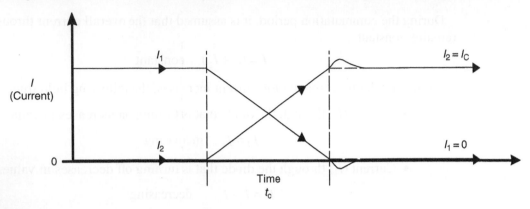

Figure 6.7
The currents in each branch during commutation

Even if the commutation time is very short, the commutation voltage of an AC-fed rectifier bridge does not remain constant but changes slightly during the commutation period. An increasing commutation voltage will tend to reduce the commutation time.

6.5.2   Three-phase commutation with six-diode bridge

In practical power electronic converter circuits, the commutation follows the same basic sequence outlined above. Figure 6.8 shows a typical six-pulse rectifier bridge circuit to convert three-phase AC currents I_A, I_B, and I_C, to a DC current I_D.

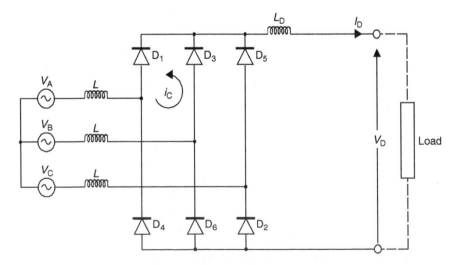

Figure 6.8
Three-phase commutation with a six-pulse diode bridge

This type of circuit is relatively simple to analyze because only two of the six diodes conduct current at any one time. The idealized commutation circuit can easily be identified. In this example, the commutation is assumed to be taking place from diode D_1 to D_3 in the positive group, while D_2 conducts in the negative group.

In power electronic bridge circuits, it is conventional to number the diodes D_1 to D_6 in the sequence in which they are turned ON and OFF. When V_A is the highest voltage and V_C the lowest, D_1 and D_2 are conducting.

Similar to the idealized circuit in Figure 6.8, when V_B rises to exceed V_A, D_3 turns on and commutation transfers the current from diode D_1 to D_3. As before, the commutation time is dependent on the circuit inductance (L) and the commutation voltage ($V_B - V_A$). As can be seen from the six-pulse diode rectifier bridge example in Figure 6.4, the commutation is usually initiated by external changes.

In this case, the three-phase supply line voltages control the commutation. In other applications, the commutation can also be initiated or controlled by other factors, depending on the type of converter and the application. Therefore, converters are often classified in accordance with the source of the external changes that initiate commutation. In the above example, the converter is said to be line-commutated because the source of the commutation voltage is on the mains supply line. A converter is said to be self-commutated if the source of the commutation voltage comes from within the converter itself. Gate-commutated converters are typical examples of this.

6.5.3   Line-commutated diode rectifier bridge

One of the most common circuits used in power electronics is the three-phase line-commutated six-pulse rectifier bridge (Figure 6.9), which comprises of six diodes in a bridge connection. Single-phase bridges will not be covered here because their operation can be deduced as a simplification of the three-phase bridge.

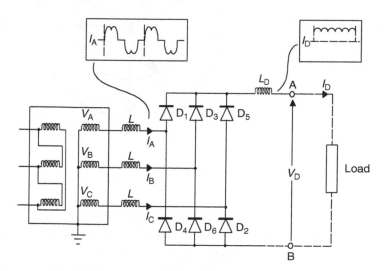

Figure 6.9
Line-commutated diode rectifier bridge

Assumptions

- The supply voltages are stiff and completely sinusoidal
- Commutations are instantaneous and have no recovery problems
- Load currents are completely smooth

- Transformers and other line components are linear and ideal
- There is no volt drop in power electronic switches.

These assumptions are made to gain an understanding of the circuits and to make estimates of currents, voltages, commutation times, etc. In addition, the limiting conditions that affect the performance of the practical converters and their deviation from the ideal conditions will be examined to bridge the gap from the ideal, to the practical.

In the diode bridge, the diodes are not controlled from an external control circuit. Instead, the commutation is initiated externally by the changes that take place in the supply line voltages, hence the name line-commutated rectifier.

According to convention, the diodes are labeled D_1 to D_6 in the sequence in which they are turned ON and OFF. This sequence follows the sequence of the supply line voltages.

The three-phase supply voltages comprise three sinusoidal voltage waveforms, 120° apart, which rise to their maximum value in the sequence A–B–C. According to convention, the phase-to-neutral voltages are labeled V_A, V_B, and V_C and the phase-to-phase voltages are V_{AB}, V_{BC} and V_{CA}, etc.

These voltages are usually shown graphically as a vector diagram, which rotates counter-clockwise at a frequency of 50 times per second. A vector diagram of these voltages and their relative positions and magnitudes are shown in Figure 6.10. The sinusoidal voltage waveforms, of the supply voltage, may be derived from the rotation of the vector diagram.

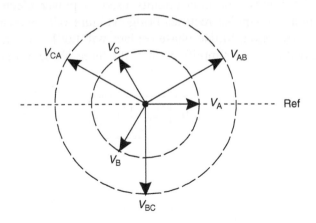

Figure 6.10
Vector diagram of the three-phase mains supply voltages

The output of the converter is the rectified DC voltage V_D, which drives a DC current I_D through a load on the DC side of the rectifier. In the idealized circuit, it is assumed that the DC current I_D is constant and completely smooth and without ripple. The bridge comprises of two commutation groups, one connected to the positive leg, consisting of diodes D_1–D_3–D_5, and one connected to the negative leg, consisting of the diodes D_4–D_6–D_2. The commutation transfers the current from one diode to another in sequence and each diode conducts the current for 120° of each cycle, as shown in Figure 6.10.

In the upper group, the positive DC terminal follows the highest voltage in the sequence V_A–V_B–V_C via diodes D_1–D_3–D_5. When V_A is near its positive peak, the diode D_1 conducts and the voltage of the positive DC terminal follows V_A. The DC current flows through the load and returns via one of the lower group diodes. With the passage of time, V_A reaches its sinusoidal peak and starts to decline. At the same time, V_B rises and eventually reaches a point, when it becomes equal to and starts to exceed V_A. At this point, the forward voltage across diode D_3 becomes positive and it starts to turn on. The commutating voltage in this circuit, V_B–V_A starts to drive an increasing commutation current though the circuit inductances and the current through D_3 start to increase, as the current in D_1 decreases. In a sequence of events similar to that described above, the commutation takes place and the current is transferred from diode D_1 to diode D_3. At the end of the commutation period, the diode D_1 is blocked and the positive DC terminal follows V_B until the next commutation takes place, to transfer the current to diode D_5. After the diode D_5, the commutation transfers the current back to D_1 and the cycle is repeated.

In the lower group, a similar sequence of events takes place, but here the negative voltages and the current flow from the load back to the mains. Initially, D_2 is assumed to be conducting when V_C is more negative than V_A. As time progresses, V_A becomes equal to V_C and then becomes more negative. The commutation takes place and the current is transferred from diode D_2 to D_4. Diode D_2 turns off and D_4 turns on. The current is later transferred to diode D_6, then back to D_2 and the cycle is repeated.

In Figure 6.10, the conducting periods of the diodes in the upper and lower groups are shown over several cycles of the three-phase supply. This shows that only two diodes conduct current at any time (except during the commutation period, which is assumed to be infinitely short) and that each of the six diodes conduct for only one portion of the cycle in a regular sequence. The commutation takes place alternatively in the top group and the lower group.

The DC output voltage V_D is not a smooth voltage and consists of portions of the phase-to-phase voltage waveforms. For every cycle of the 50 Hz AC Waveform (20 ms), the DC voltage V_D comprises portions of six voltage pulses, V_{AB}, V_{AC}, V_{BC}, V_{BA}, V_{CA}, V_{CB}, etc., hence the name, six-pulse rectifier bridge.

The average magnitude of the DC voltage may be calculated from the voltage waveform shown in Figure 6.10. The average value is obtained by integrating the voltage over one of the repeating 120° portions of the DC voltage curve. This integration yields an average magnitude of the voltage V_D as follows:

$$V_D = 1.35 \times (\text{RMS phase} - \text{Phase voltage})$$
$$V_D = 1.35 \times V_{RMS}$$

For example, if $V_{RMS} = 415$ V, then $V_D = 560$ V DC. When there is a sufficient inductance in the DC circuit, then DC current I_D will be steady and the AC supply current will comprise of segments of DC current from each diode in sequence.

As an example, the current in the A-phase is shown in Figure 6.11. The non-sinusoidal current that flows in each phase of the supply mains can affect the performance of any other AC equipment connected. In practice, to ensure that the diode reverse blocking voltage capability is properly specified, it is necessary to know the magnitude of the reverse blocking voltage that appears across each of the diodes to the supply line that is designed to operate with sinusoidal waveforms.

Theoretically, the maximum reverse voltage across a diode is equal to the peak of the phase–phase voltage. For example, the reverse voltage V_{CA} and V_{CB} appear across diode D_5 during the blocking period. In practice, a safety factor of 2.5 is commonly used for specifying the reverse-blocking capability of diodes and other power electronic switches. On a rectifier bridge fed from a 415 V power supply, the reverse blocking voltage V_{bb} of the diode must be higher than 2.5×440 V $= 1100$ V. Therefore, it is common practice to use diodes with a reverse-blocking voltage of 1200 V.

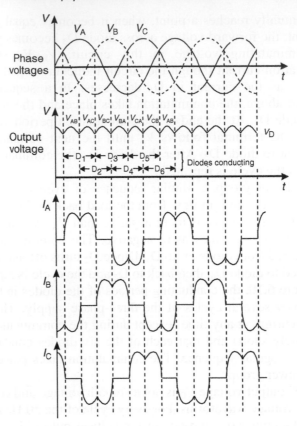

Figure 6.11
Voltage and current waveforms during commutation

6.5.4   The line-commutated thyristor rectifier bridge

The output DC voltage and the operational sequence of the diode rectifier in Figure 6.9, is dependent on the continuous changes in the supply line voltages and is not dependent on any control circuit. Therefore, it is called an uncontrolled diode rectifier bridge because the DC voltage output is uncontrolled and is fixed at $1.35 \times V_{RMS}$.

If the diodes are replaced with thyristors, it then becomes possible to control the point at which the thyristors are triggered and therefore, the magnitude of the DC output voltage can be controlled. This type of a converter is called a controlled thyristor rectifier bridge. This requires an additional control circuit, to trigger the thyristor, at the right instant. A typical six-pulse thyristor converter is shown in Figure 6.12.

Based on the previous chapter, the conditions required for a thyristor to conduct current in a power electronic circuit are given below:

- A forward voltage must exist across the thyristor
- A positive pulse must be applied to the thyristor gate.

If each thyristor were triggered at the instant when the forward voltage across it tends to become positive, then the thyristor rectifier operates in the same way as the diode rectifier described above. All voltage and current waveforms of the diode bridge apply to the thyristor bridge.

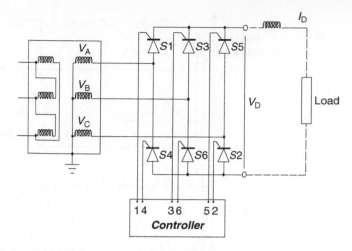

Figure 6.12
Six-pulse controlled thyristor rectifier bridge

A thyristor bridge operating in this mode is said to be operating with a zero delay angle and gives a voltage output of:

$$V_D = 1.35 \times V_{RMS}$$

The output of the rectifier bridge can be controlled, by delaying the instant at which the thyristor receives a triggering pulse. This delay is usually measured in degrees, from the point at which the switch CAN turn on, due to the forward voltage becoming positive. The angle of delay is called the delay angle, or sometimes the firing angle, and is designated by the symbol α. The reference point, for the angle of delay, is the point where a phase voltage wave crosses the voltage of the previous phase and becomes positive, relative to it. A diode rectifier can be thought of as a converter with a delay angle of $\alpha = 0°$. The main purpose of controlling a converter is to control the magnitude of the DC output voltage. In general, the larger the delay angle, the lower the average magnitude of the DC voltage. Under the steady state operation, of a controlled thyristor converter, the delay angle for each switch is the same. Figure 6.13 shows the voltage waveforms, where the triggering of the switches has been delayed by an angle of α degrees.

Operation

In the positive switch group, the positive DC terminal follows the voltage associated with the switch, which is in conduction in the sequence V_A–V_B–V_C. Assume initially, that thyristor S1 associated with voltage V_A is conducting and S3 is not yet triggered. The voltage on the positive bus on the DC side follows the declining voltage V_A because, in the absence of an S3 conduction, there is still a forward voltage across S1 and it will continue to conduct.

When an S3 is triggered after a delay angle = α, the voltage on the positive bus jumps to V_B, whose value it then starts to follow. At this instant, with both S1 and S3 conducting, a negative commutation voltage equal to V_B–V_A appears across the switch S1 for the commutation period, which then starts to turn off. With the passage of time, V_B reaches its sinusoidal peak and starts to decline, followed by the positive DC terminal. Simultaneously, V_C rises and when S5 is triggered, the same sequence of events is repeated and the current is commutated to S5.

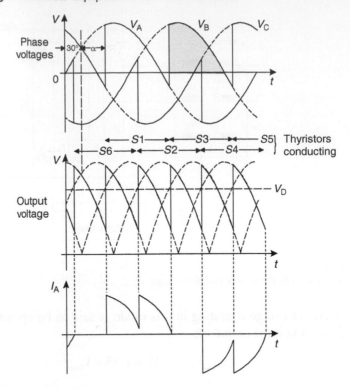

Figure 6.13
Voltage waveforms of a controlled rectifier

As with the diode rectifier, the average magnitude of the DC voltage V_D can be calculated, by integrating the voltage waveform over a 120° period, representing a repeating portion of the DC voltage. At a delay angle α, the DC voltage is given by the following equation:

$$V_D = 1.35 \times (\text{RMS phase} - \text{Phase voltage}) \times \cos\alpha$$

$$V_D = 1.35 \times V_{RMS} \times \cos\alpha$$

This formula shows that the theoretical DC voltage output of the thyristor rectifier with a firing angle $\alpha = 0$ is the same as that for a diode rectifier. It also shows that the average value of the DC voltage will decrease as the delay angle is increased and is dependant on the cosine of the delay angle. When $\alpha = 90°$, then $\cos \alpha = 0$ and $V_D = 0$, which means that the average value of the DC voltage is zero. The instantaneous value of the DC voltage is a saw-tooth voltage, as shown in Figure 6.14.

If the delay angle is further increased, the average value of the DC voltage becomes negative. In this mode of operation, the converter operates as an inverter. It is interesting to note that the direction of the DC current remains unchanged because the current can only flow through the switches in one direction. However, with a negative DC voltage, the direction of the power flow is reversed, and the power flows from the DC side to the AC side. A steady state operation, in this mode, is only possible, if there is a voltage source on the DC side. The instantaneous value of the DC voltage for $\alpha > 90°$ is shown in Figure 6.15.

In practice, the commutation is not instantaneous and lasts for a period dependant on the circuit inductance and the magnitude of the commutation voltage. As in the idealized case, it is possible to estimate the commutation time, from the commutation circuit inductance and an estimate of the average commutation voltage.

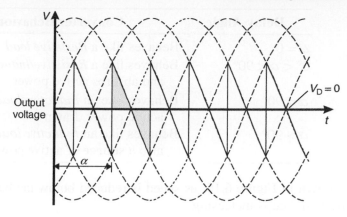

Figure 6.14
DC output voltage for delay angle $\alpha = 90°$

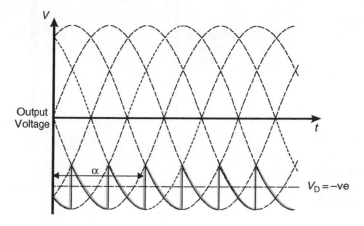

Figure 6.15
DC voltage when the delay angle $\alpha > 90°$

As in the diode rectifier, the steady DC current I_D comprises segments of current from each of the three phases on the AC side. On the AC side, the current in each phase comprises of non-sinusoidal blocks, similar to those associated with the diode rectifier and with similar harmonic consequences. In the case of the diode bridge, with a delay angle of $\alpha = 0$, the angle between the phase current and the corresponding phase voltage on the AC side is approximately zero. Consequently, the power factor is unity and the converter behaves like a resistive load.

For the controlled rectifier, with a delay angle of α, the angle between the phase current and the corresponding phase voltage is also α, and called the power factor angle ϕ. This angle should be called the displacement factor because it does not really represent the power factor. Consequently, when the delay angle of the thyristor rectifier is changed to reduce the DC voltage, the angle between the phase current and voltage also changes by the same amount. The converter then behaves like a resistive-inductive load with a displacement factor of $\cos\phi$. It is well known that the power factor associated with a controlled rectifier falls, when the DC output voltage is reduced.

A common example of this is a DC motor drive controlled by a thyristor converter. As the DC voltage is reduced, to reduce the DC motor speed, at a constant torque, the power factor drops and more reactive power is required at the supply line to the converter.

Delay Angle	Converter Behavior
$\alpha = 0°$	Behaves like a *Resistive load*
$0° < \alpha < 90°$	Behaves like a *Resistive/Inductive load* and absorbs active power
$\alpha = 90°$	Behaves like an *Inductive load* with no active power drawn
$\alpha > 90°$	Behaves like an *Inductive load* but is also a **source** of active power

As shown in Figure 6.16, as speed is reduced below the base speed, the reactive power requirement keeps increasing.

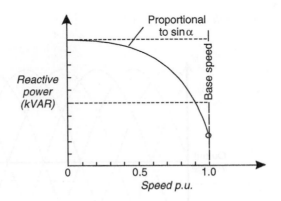

Figure 6.16
Reactive power requirements of a DC motor drive with a constant torque load fed from a line-commutated converter

Practical limitations of line-commutated converters

The above analysis covers the theoretical aspects of both uncontrolled and controlled converters. In practice, the components are not ideal and the commutations are not instantaneous. This results in certain deviations from the theoretical performance. One of the deviations is that the DC load current is never completely smooth.

Reasons

- Accepting that the instantaneous DC voltage V_D can never be completely smooth, if the load is purely resistive, the DC load current cannot be completely smooth because it will linearly follow the DC voltage.
- Also, at delay angles $\alpha > 60°$, the DC output voltage becomes discontinuous and, consequently, so would the DC current.

Remedy

In an effort to maintain a smooth DC current, practical converters usually have some inductance L_D in series with the load on the DC side. For complete smoothing, the value of L_D should theoretically be infinite, which is not practical.

The practical consequence of this is that the theoretical formula for the calculated value of DC voltage ($V_D = 1.35\ V_{RMS} \cos \alpha$) is not completely true for all values of the delay angle α. Practical measurements confirm that it only holds true for delay angles of up to 75°, but this depends on the load type and in particular, the DC load inductance. Experience shows that for a particular delay angle $\alpha > 60°$, the average DC voltage will be higher than the theoretical value, as shown in the Figure 6.17.

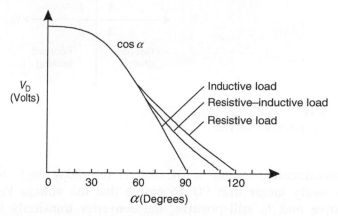

Figure 6.17
Deviation of DC voltage from theoretical vs delay angle

Applications for line-commutated rectifiers

An important application of the line-commutated converter is the DC motor drive. Figure 6.18 shows a single controlled line-commutated converter connected to the armature of a DC motor. The converter provides a variable DC voltage V_A to the armature of the motor. This is how the control circuit of the converter is used to change the motor speed.

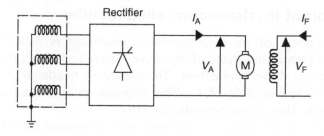

Figure 6.18
Converter-fed DC motor drive

When the delay angle is less than 90°, the DC voltage is positive and a positive current I_A flows into the armature of the DC motor, to deliver active power to the load. The drive system is said to be operating in the 1st quadrant (Figure 6.19), where the motor runs in the forward direction, with a transfer of active power from the supply to the motor and its mechanical load.

The motor field winding is separately excited from a simple diode rectifier and carries a field magnetizing current I_F. For a fixed field current, the speed of the motor is proportional to the DC voltage at the armature. The speed can be controlled by varying the delay angle of the converter and its output armature voltage V_A.

If the delay angle of the converter is increased to an angle greater than 90°, the voltage V_D will become negative and the motor will slow to a standstill. The current I_D also reduces to zero and the supply line can be disconnected from the motor without breaking any current.

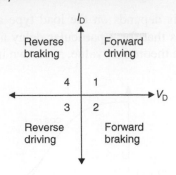

Figure 6.19
Operating quadrants for VSD

Consequently, to stop a DC motor, the delay angle must be increased to a value sufficiently larger than 90° to ensure that the voltage V_D becomes negative. With V_D negative and I_D still positive, the converter transiently behaves like a generator and produces a braking torque.

In addition, this acts as a brake to slow the motor and its load quickly to a standstill. In this situation, the drive system is said to be operating in the 2nd quadrant where the motor is running in the forward direction. The converters discussed so far have been single converters, which can only operate with a positive DC current (I_D = +ve), which means that the motor can only run in the forward direction but an active power can be transferred in either direction. Single DC converters can only operate in quadrants 1 and 4 and are known as second quadrant converters.

6.5.5   Quadrant thyristor-controlled rectifier

The concept of the four operating quadrants is illustrated below. It shows the four possible operating states of any drive system and shows the directions of V_D and I_D for the DC motor drive application. To operate in quadrants 3 and 2, it must be possible to reverse the direction of I_D. This requires an additional converter bridge connected for current to flow in the opposite direction.

This type of converter is known as a four-quadrant DC converter, and is sometimes called a double or back-to-back six-pulse rectifier (Figure 6.20).

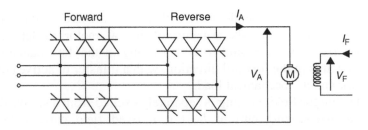

Figure 6.20
Four-quadrant line-commutated rectifier

With a DC motor drive fed from a four-quadrant DC converter, the operation in all four quadrants is possible with a speed control in either the forward or reverse direction.

Operation

A change of direction of the motor can quickly be achieved. Converter-1 is used as a controlled rectifier for speed control in the forward direction of rotation, while converter-2 is blocked, and vice versa in the reverse direction.

Assume initially, that the motor runs in the forward direction, under the control of Converter-1, with a delay angle of <90° and Converter-2 is blocked. The changeover sequence from running in the forward direction to the reverse direction is as follows:

- Converter-1 delay angle increased to $\alpha > 90°$. This means that DC voltage $V_D < 0$ and DC current I_D is decreasing.
- When $I_D = 0$, Converter-1 is blocked and thyristor firing is terminated.
- After a small delay, Converter-2 is unblocked and starts in the inverter mode with a firing angle greater than 90°.
- If the motor is still turning in the forward direction, converter-2 DC current I_D starts to increase in the negative direction and the DC machine acts as a generator and is broken to standstill, returning energy to the supply line.
- As the firing angle is reduced to $\alpha < 90°$, converter-2 changes from the inverter to the rectifier mode, and as voltage V_D increases, the motor starts to rotate in the opposite direction.

In a DC motor drive, the reversal rotation direction can also be achieved by using a single converter and by changing the direction of the excitation current.

This method can only be used where there are no special drive requirements for changing over from the forward to the reverse operation. In this case, using switches in the field circuit do the changeover mechanically during a period at standstill. Considerable time delays are required during a standstill, to demagnetize the field in the reverse direction. There are many practical applications for both the uncontrolled and the controlled line-commutated rectifiers. Some of the more common applications include the following:

- DC motor drives with variable speed control
- DC supply for variable voltage-variable frequency inverters
- Slip-energy recovery converters for wound rotor induction motors
- DC excitation supply for machines
- High-voltage DC converters.

6.6   Gate-commutated inverters (DC/AC converters)

Most modern AC VSDs in the 1–500 kW range are based on Gate-commutated devices such as the GTO, MOSFET, BJT, and IGBT, which can be turned ON and OFF by low-power control circuits connected to their control gates.

Operating principle   An inverter works on a DC supply giving a variable frequency AC output. It can be operated either as a step wave inverter or a PWM inverter.

In a step wave inverter, the transistors are switched such that the phase difference is 60° and each transistor is kept on for 180°. To vary the output AC waveform frequency, the duration between the turn on of transistors is changed. The output of AC voltage is varied by changing the DC input voltage. This type of an inverter has problems of a pulsating torque due to harmonics in the output voltage. This gives a pulsating motion of the rotor at low speeds.

The pulsating torque can be eliminated by the use of pulse width modulation (PWM) type inverters since their output has low harmonic content. The details of this type of inverter are explained later in this chapter. With a DC supply and semiconductor power electronic switches, it is not possible to obtain a pure sinusoidal voltage at the load. On the other hand, it may be possible to generate a near-sinusoidal current. Consequently, the objective is that the current through the inductive circuit should approximate a sinusoidal current as closely as possible.

6.6.1   Single-phase square wave inverter

To establish the principles of gate-controlled inverter circuits, Figure 6.21 shows four semiconductor power switches feeding an inductive load from a single-phase supply.

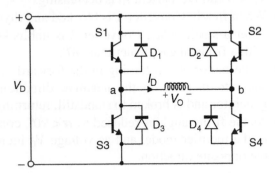

Figure 6.21
Single-phase DC to AC inverter

This circuit can be considered as an electronic reversing switch, which allows the input DC voltage V_D to be connected to the inductive load in any one of the following ways:

1. S1 = on, S4 = on.... giving $+ V_D$ at the load
2. S2 = on, S3 = on.... giving $- V_D$ at the load
3. S1 = on, S2 = on.... giving zero volts at the load
4. S3 = on, S4 = on.... giving zero volts at the load
5. S1 = on, S3 = on.... giving a short-circuit fault
6. S2 = on, S4 = on.... giving a short-circuit fault.

However, these four switches can be controlled to give a square waveform across the inductive load, as shown in Figure 6.20. This makes use of the switch configurations (1) and (2), but not the switch configuration (3) or (4). Clearly, for a continued safe operation, option (4) should always be avoided. In the case of a purely inductive load, the current waveform is a triangular waveform, as shown in Figure 6.22. In the first part of the cycle, the current is negative although only switches S1 and S4 are on. Since most power electronic devices cannot conduct negatively, to avoid damage to the switches, this negative current would have to be diverted around them.

Consequently, diodes are usually provided, anti-parallel with the switches to allow the current flow to continue. These diodes are sometimes called reactive or freewheeling diodes. These diodes conduct when the voltage and current polarities are opposite. This occurs when there is a reverse power flow back to the DC supply.

The frequency of the periodic square wave output is called the fundamental frequency. Using Fourier analysis, any repetitive waveform can be resolved into a number of sinusoidal waveforms. Each comprises one sinusoid at a fundamental frequency and a number of sinusoidal harmonics at higher frequencies, which are multiples of the fundamental frequency. The harmonic spectrum for a single-phase square wave output is shown in Figure 6.23. With an increase in frequency, the amplitude of the higher-order harmonics voltages fall off rapidly.

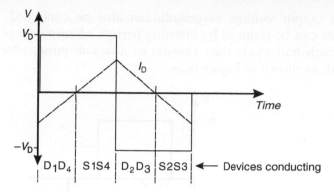

Figure 6.22
Square wave modulation waveforms

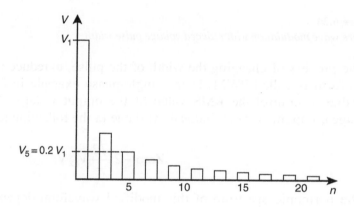

Figure 6.23
Square-wave harmonic spectrum

The RMS value of the fundamental sinusoidal voltage component is:

$$V_1 = 2 \frac{\sqrt{2}}{\pi} V_d \text{ V}$$

The RMS value of the nth harmonic voltage:

$$V_n = \frac{V_1}{n} \text{ V}$$

This illustrates that the square wave output voltage, has many unwanted components of reasonably large magnitude at frequencies close to the fundamental. The current flow in the load is due to an output voltage distortion, as demonstrated by the non-sinusoidal current wave-shape. In this example, the current has a triangular shape.

If the square-wave voltage were presented to a single-phase induction motor, the motor would run at the frequency of the square-wave. Being a linear device (inductive/resistive load), however, it would draw non-sinusoidal currents and would suffer additional heating due to the harmonic currents. These currents may also produce pulsating torques.

To change the speed of the motor, the fundamental frequency of the inverter output can be changed by adjusting the switching speed. To increase frequency, the switching speed can be increased, and to decrease frequency, the switching speed can be decreased.

The output voltage magnitude can also be controlled. The average inverter output voltage can be reduced by inserting periods of zero voltage, using a switch configuration (3). Each half cycle then consists of a square pulse, which is only a portion of a half period, as shown in Figure 6.24.

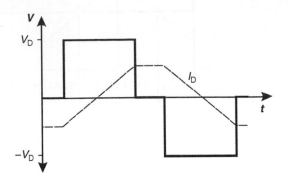

Figure 6.24
Square wave modulation with reduced voltage pulse width

The process of changing the width of the pulse, to reduce the average RMS value of a waveform is called PWM. In the single-phase example in Figure 6.24, PWM makes it possible to control the RMS value of the output voltage. The fundamental sinusoidal voltage component is continuously variable in the following range:

$$\text{Zero} - 2 \frac{\sqrt{2}}{\pi} V_D \text{ V}$$

The harmonic spectrum of this modified waveform depends on the fraction, that the pulse is, of the full square wave, but is broadly similar to the waveform shown earlier.

6.6.2 Single-phase pulse width modulation (PWM) inverter

The fact that the voltage supply to the stator, of an AC induction motor, is a square wave and is not distorted in itself is a problem to the motor. The main problem comes from the distortion of the current waveform, which results in extra copper losses and is due to shaft torque pulsations. The ideal inverter output is one, which results in a current waveform of low harmonic distortion.

An AC induction motor is predominantly inductive, with a reactance that depends on the frequency ($X_L = j2\pi f L$). It is, therefore, beneficial if the voltage harmonic distortion can be pushed into high frequencies, where the motor impedance is high and not much distorted current will flow.

One technique for achieving this is the sine-coded pulse width modulation (sine-PWM). This requires the power devices to be switched, at frequencies much greater than that of the fundamental frequency, producing a number of pulses, for each portion of the desired output period. The frequency of the pulses is called modulation frequency.

The width of the pulses is varied throughout the cycle in a sinusoidal manner, giving a voltage waveform as shown in Figure 6.25. The figure also shows the current waveform for an inductive load, by showing the improvement in a waveform.

The improvement in the current waveform can be explained by the harmonic spectrum shown in Figure 6.26. It can be seen that, although the voltage waveform still has many distortion components, they now occur at higher harmonic frequencies, where the high-load impedance of the motor is effective in reducing these currents.

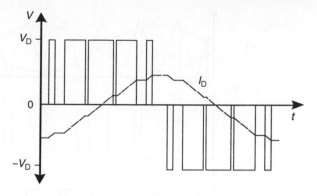

Figure 6.25
Sine-coded PWM voltage and current

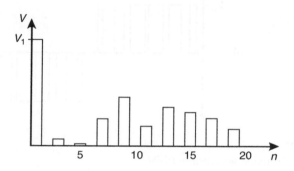

Figure 6.26
Harmonic spectrum for a PWM inverter

Increasing the modulation frequency will improve the current waveform, but at the expense of increased losses in the switching devices of the inverter. The choice of modulation frequency depends on the type of switching device and its frequency. With the force-commutated thyristor inverter, a modulation frequency of up to 1 kHz was possible with the older technologies.

With the introduction of GTOs and BJTs, this could be pushed up to around 5 kHz. With IGBTs, the modulation frequency could be as high as 20 kHz.

In practice, a maximum modulation frequency of up to 12 kHz is common with IGBT inverters up to about the 22 kW motor size and 8 kHz for motors up to about 500 kW. The choice of modulation frequency is a trade-off, between the losses in the motor and in the inverter. At low-modulation frequencies, the losses in the inverter are low and those in the motor are high. At high-modulation frequencies, the losses in the inverter increase, while those in the motor decrease.

One of the most common techniques for achieving the sine-coded PWM in practical inverters is the sine–triangle intersection method shown in Figure 6.27. A triangular saw-tooth waveform is produced in the control circuit at the desired inverter-switching frequency. This is compared in a comparator, with a sinusoidal reference signal, which is equal in frequency and proportional in magnitude to that of the desired sinusoidal output voltage. The voltage V_{AN} (Figure 6.27(b)) is switched high whenever the reference waveform is greater than the triangle waveform. The voltage V_{BN} (Figure 6.27(c)) is not controlled by the same triangle waveform but with a reference waveform shifted by 180°.

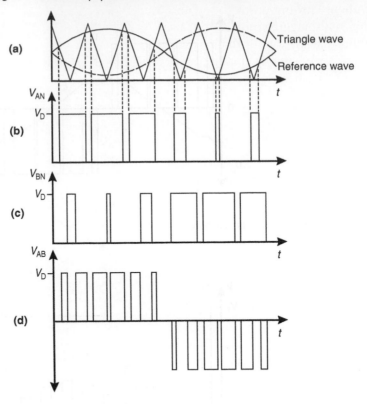

Figure 6.27
Principle of triangle intersection PWM

The actual phase-to-phase output voltage is then V_{AB} (Figure 6.27(d)), which is the difference between V_{AN} and V_{BN}, which consists of a series of pulses, each of whose width is related to the value of the reference sine wave at that time.

The number of pulses in the output voltage V_{AB} is double that in the inverter leg voltage V_{AN}. For example, an inverter switching at 5 kHz should produce a switching distortion at 10 kHz in the output phase-to-phase voltage. The polarity of the voltage is alternatively positive and negative at the desired output frequency.

It can also be seen that the reference sine wave in Figure 6.27 is given a DC component so that the pulse produced by this technique has a positive width. This puts a DC bias on the voltage of each leg as shown in Figures 6.2(b) and (c). However, each leg has the same DC offset which disappears from the load voltage.

The technique of using a sine-triangle intersection is particularly suited to the old analog control circuits, where the two reference waveforms are fed into a comparator and the output of the comparator is used to trigger the inverter switches. Modern digital techniques operate based on a switching algorithm. For example, by producing triggering pulses proportional to the area under a part of the sine wave.

Recently, manufacturers have developed a number of different algorithms that optimize the performance of the output waveforms for AC induction motors. These techniques result in PWM output waveforms, which are similar to those shown in Figure 6.27. The sine-coded PWM voltage waveform is a composite of a high-frequency square wave at the pulse frequency (the switching carrier) and the sinusoidal variation of its width (the modulating waveform). It has been found that, for the lowest harmonic distortion, the modulating waveform should be synchronized with the carrier frequency, so it contains an integral number of carrier periods.

This requirement becomes less important with high carrier frequencies of more than twenty times the modulating frequency. The voltage and frequency of a sinusoidal PWM waveform are varied by changing the reference waveform of Figure 6.27(a), giving outputs as shown in Figure 6.28.

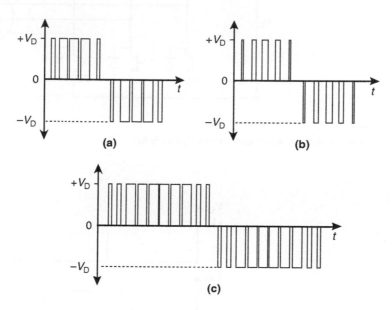

Figure 6.28
Variation of frequency and voltage with sinusoidal PWM

Figure 6.28(a) shows a base case, with the rated *V/f* ratio. Figure 6.28(b) shows the case where the voltage reference is halved, resulting in the halving of each pulse. Figure 6.28(c) shows the case where the reference frequency is halved, resulting in the extension of the modulation over twice as many pulses.

The largest voltage with the sine-coded PWM occurs when the pulses in the middle are the widest, giving an output with a peak voltage equal to the supply.

Modulation index

This is defined as the ratio of the peak AC output to the DC supply. Thus, the largest output voltage occurs when the modulation index is 1.

It is possible to achieve a high value of modulation index by abandoning the strict sine-PWM and by adding some distortion to the sinusoidal reference voltage. This results in the removal of some of the pulses near the center of the positive and negative parts of the waveform. This is a process called pulse dropping. In the limit, a square voltage waveform can be achieved with a modulation index of 1.

6.6.3   Three-phase inverter

A three-phase inverter could be constructed from three inverters of the type shown earlier. However, it is more economical to use a six-pulse (three-leg) bridge inverter as shown in Figure 6.29.

In its simple form, a square output voltage waveform can be obtained by switching each leg high for one half-period and low for the next half-period, at the same time ensuring that each phase is shifted one third of a period (120°), as shown in the Figure 6.30.

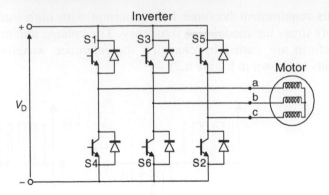

Figure 6.29
Three-phase inverter using gate-controlled switches

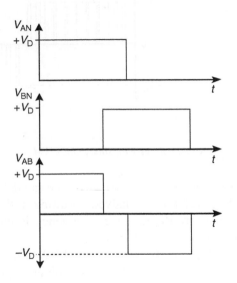

Figure 6.30
Quasi-Square wave modulation output waveforms

The resulting phase-to-phase voltage waveform comprises a series of square pulses whose widths are two-thirds of the period of the switch, in each phase.

The resulting voltage waveform is called a quasi-square wave (QSW) voltage. This simple technique was used in early voltage source inverters (VSI), which used the forced commutated thyristors in the inverter bridge. To maintain a constant *V/f* ratio, the rectifier bridge controlled the magnitude of the DC bus voltage, in order to keep a fixed ratio to the output frequency, which was controlled by the inverter bridge. This technique was also known as the pulse amplitude modulation (PAM).

The output voltage of a three-phase converter has a harmonic spectrum, very similar to the single-phase square wave, except that the triplen harmonics (harmonics whose frequency is a multiple of three times the fundamental frequency) have been eliminated. In an inverter with a three-phase output, this means that the 3rd, 9th, 15th, 21st, etc. harmonics are eliminated. To develop a three-phase variable voltage AC output of a particular frequency, the voltages V_{AN}, V_{BN}, V_{CN} on the three output terminals a, b, and c in Figure 6.29 can be modulated on and off to control both the voltage and the frequency.

The pulse–width ratio over the period can be changed according to a sine-coded PWM algorithm (Figure 6.31).

When the phase–phase voltage V_{AB} is formed, the present modulation strategy gives only positive pulses for a half period followed by negative pulses for a half period, a condition known as consistent pulse polarity. It can be shown that the consistent pulse polarity guarantees the lowest harmonic distortion, with most of the distortion being at twice the inverter chopping frequency. Therefore, these are the types of inverters used in industrial applications. The same methods are also used in AC drives.

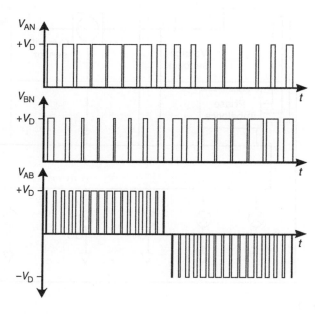

Figure 6.31
Output voltage waveform of a three-phase sine-coded PWM

6.7   Overall protection and diagnostics

Figure 6.32 is a summary of the protection features commonly used in modern digital PWM AC converters. As outlined above, many of these protection functions are implemented in software, using suitable algorithms. The main exceptions are the over-current protection and the earth fault protection. These are implemented in hardware, to ensure that they are sufficiently fast to adequately protect the power semiconductor devices.

The normal protection features that are available with an inverter-based VSD are phase imbalance, under voltage, over voltage, over current, over temperature, and earth fault. Protection features require, sensor-like CTs, voltage transformers, and temperature sensors. These are installed and adequate indications are provided for the same.

6.7.1   Operator information and fault diagnostics

Modern digital VSDs all have some form of an operator interface module. This module provides access to internal data, about the control and status parameters during normal operation and diagnostic information during fault conditions. This module is called the human interface module (HIM) or a name on similar lines. The HIM usually provides an

LED or LCD display and buttons to interrogate the control circuit. This operator interface can also be used to install and change the VSD settings parameters.

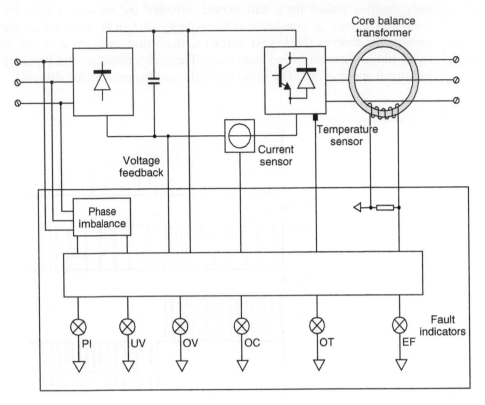

Figure 6.32
Example of VSD protection block diagram

When an internal or an external fault occurs, the control circuit registers the fault type. This helps to identify the cause of the fault and the subsequent rectification required. Modern microprocessor-controlled converters employ a diagnostic system. The system monitors both, the internal and the external operating conditions and responds to any faults. This is done in the manner programed by the user. The control system retains the fault information in a non-volatile memory for later analysis. This feature is known as fault diagnostics.

There are three main levels of operator information and fault diagnostics in reputed models, as given below:

- The first level provides information about the on-going situation inside a VSD. It refers mainly to the setting parameters and the real-time operating parameters. It meters information, such as output voltage, output current, output frequency, etc.
- The second level provides diagnostic information, about the status of the protection circuits, and indicates the external faults, as described above.
- The third level provides diagnostic information about the status of internal faults, such as the identification of failed modules. Dedicated internal diagnostics are usually only found in high-performance VSDs.

Figure 6.33 is a brief list of typical internal parameters and fault conditions.

Module	Parameters and Fault Diagnostics
Power supply	Power supply voltage, current, and frequency
DC bus	DC link voltage and current
Motor	Output voltage, current, frequency, speed, torque, temperature
Control signals	Set point, process variable, error, ramp times
Status	Protection circuits, module failures, internal temps, fans running, switching frequency, current limit, motor protection, etc.
Fault conditions	Power device fault, power supply failed, driver circuit failed, current feedback failed, voltage feedback failed, main controller failed

Figure 6.33
Typical list of VSD parameters

At the first level, most modern digital VSDs provide information about the status of the following:

- All setting parameters which define operating conditions
- The digital inputs (DI) and digital outputs (DO), such as start, stop, enabled, jog, forward/reverse, etc.
- The status of the analog inputs (AI), such as speed reference, torque reference, etc.
- The real-time operating parameters, which include a vast array of information, such as output frequency, output voltage, output current, etc.

At the second level, when a fault occurs and the VSD stops, diagnostic information is provided to assist in the rectification of the fault, thereby reducing downtime. There is always some overlap between these levels of diagnostics. For example, a persistent over-current trip with no motor connection can indicate a failed power electronic switching device inside the converter.

Figure 6.34 shows the most common external fault indications provided by the VSD diagnostics system, and the possible internal or external problems which may have caused them.

Protection	Internal Fault	External Fault
Over-voltage	Deceleration rate set too fast	Mains voltage too high Transient over-voltage spike
Under-voltage	Internal power supply failed	Mains voltage too low Voltage sag present
Over-current	Power electronic switch failed Driver circuit failed	Short-circuit in motor or cable
Thermal overload	Control circuit failed	Motor overloaded or stalled
Earth fault	Internal earth fault	Earth fault in motor or cable
Over temperature	Cooling fan failed Heat sink blocked	Ambient too high Enclosure cooling blocked
Thermistor trip		Motor thermistor protection

Figure 6.34
VSD diagnostics table

The internal diagnostics system can provide an operator with information about faults that have occurred inside the drive. This can be further broken down into fault conditions, such as a failed output device, commutation failures, etc. Fault conditions are indications that a particular module or device has failed or is not operating normally.

To provide fault condition monitoring, the drive must be specifically designed to include internal fault diagnostic circuits. For example, power semiconductor drivers may include circuits that measure the saturation voltage, which is the voltage across the device when it is on, for each power semiconductor. This can identify a short-circuit in the power switch and the VSD can be shut down before the external over-current trips or the fuses can operate.

Considerable cost and effort is required to implement the internal fault condition monitoring. Only a few high-performance VSDs provide extensive internal diagnostics. This feature can be very useful for troubleshooting, but this is usually only warranted when downtime represents a major cost to the user.

6.8   Installations and commissioning

6.8.1   General installation guidelines

Environmental requirements

Modern power electronic AC VVVF converters, which are used for the speed control of electric motors, are usually supplied as stand-alone units in one of the following configurations. The first two are the most common configurations:

- *IP00 rating*: Designed for chassis mounting into the user's own enclosure, usually as part of a MCC.
- *IP20/IP30 rating*: Designed for mounting within a 'clean environment', such as a weatherproof, air-conditioned equipment room. The environment should be free of dust, moisture, and contaminants. The temperature should be kept within specified limits.
- *IP54 rating*: Designed for mounting outside in a partially sheltered environment, which may be dusty and/or wet.

Environmental conditions for installation

The main advantage of an AC VSD is that the TEFC squirrel-cage motor is inherently well protected from poor environmental conditions and is usually rated at IP54 or better. It can be reliably used in dusty and wet environments.

On the other hand, the AC converter is far more sensitive to its environment and should be located in an environment that is protected from the following factors:

- Dust and other abrasive materials
- Corrosive gases and liquids
- Flammable gases and liquids
- High levels of atmospheric moisture.

When installing an AC converter, the following environmental limits should be considered:

- *Specified ambient temperature*: $\leq 40\ ^\circ C$
- *Specified altitude*: ≤ 1000 m above sea level
- *Relative humidity*: $\leq 95\%$.

De-rating for high temperature

In regions or environments where there is a high ambient temperature, exceeding the accepted 40 °C specified in the standards, both the motor and the converter need to be de-rated. This means that they can only be run, at loads that are less than their 40 °C rating, to avoid thermal damage to the insulation materials.

The manufacturers of AC converters usually provide de-rating tables, for high temperature environments that are above 40 °C. The typical graph in Figure 6.35 is for a modern PWM converter. This table should be used as a guide only and should NOT be taken to apply to AC converters in general, or to any converter in particular. The design of AC converters differs from manufacturer to manufacturer. The cooling requirements are never the same. The cooling requirements, of different models, from the same manufacturer may also be different.

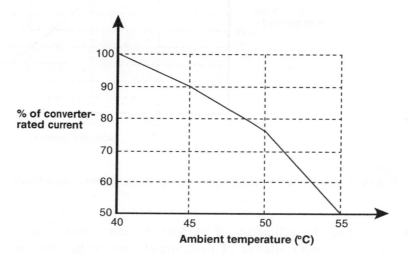

Figure 6.35
Typical temperature de-rating chart for PWM converter

De-rating for high altitude

At high altitudes, the cooling of electrical equipment is degraded, by the reduced ability of the rarified atmosphere, to remove heat from the motor or the heat sink of the converter. The reason is because the air pressure falls with an increased altitude, air density falls, and consequently, its thermal capacity is reduced. In accordance with the standards, AC converters are rated for altitudes up to 1000 m above sea level. The rated output should be de-rated for altitudes above that.

The manufacturers of AC converters usually provide the de-rating tables for altitudes higher than 1000 m. A typical table is given in Figure 6.36, for a modern IGBT-type AC converter. This table is NOT applicable to all AC converters. The de-rating of converters with high losses, such as those using BJTs or GTOs, will be much higher, than the de-rating required for low-loss IGBT or MOSFET converters. The high efficiency of the latter requires less cooling and it is less affected by altitude changes.

Mounting and enclosures for AC converters

If environmental conditions are likely to exceed these accepted working ranges, then arrangements should be made to provide additional cooling and/or environmental protection for the AC converter. The temperature limits of an AC converter are far more critical than those required for an electric motor. The temperature de-rating needs to be strictly applied. However, it is unlikely that a modern PWM converter will be destroyed if the temperature limits are exceeded. Modern AC converters have built-in thermal protection, usually a silicon junction device, mounted on the heat sink. The main problem of overtemperature tripping is associated with nuisance tripping and the associated downtime.

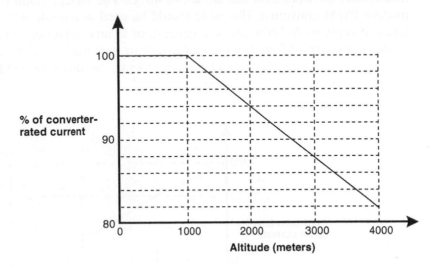

Figure 6.36
Altitude de-rating chart for IGBT-type converter (Compliments of Allen-Bradley)

Although the efficiency of modern AC converters is high, typically ±97%, they all generate a small amount of heat, mainly due to the commutation losses in the power electronic circuits. The level of losses depends on the design of the converter, the PWM switching frequency and the overall power rating. Manufacturers provide figures for the losses (Watts) when the converter is running at a full load. Adequate provision should be made to dissipate this heat into the external environment and to avoid the temperature inside the converter enclosure rising to unacceptably high levels.

Converters are air-cooled either by convection (small power ratings) or assisted by cooling fans on larger power ratings. Any obstruction to the cooling air intake and from the exhaust vents will reduce the efficiency of the cooling. The volume of air required for cooling and the power loss dissipation determines the air-conditioning requirements for the equipment room.

The cooling is also dependent on there being a temperature differential between the heat sink and the cooling air. The higher the ambient temperature, the less effective is the cooling. Both the AC converter and the motor are rated for operation in an environment where the temperature does not exceed 40 °C. When AC converters are mounted inside enclosures, care should be taken to ensure that the air temperature inside the enclosure remains within the specified temperature limits. If not, the converters should be de-rated in accordance to the manufacturer's derating tables.

In an environment where condensation is likely to occur, during the periods when the drive is not in use, anti-condensation heaters can be installed inside the enclosure. The control circuit should be designed to switch the heater on when the drive is de-energized. The heater maintains a warm dry environment inside the enclosure and avoids moisture being drawn into the enclosure when the converter is switched off and cools down.

AC converters are usually designed for mounting in a vertical position, to assist convectional cooling. On larger VSDs, cooling is assisted by one or more fans mounted at the bottom or top of the heat sink.

Many modern converters allow the following two alternative mounting arrangements:

- *Surface mounting*: The back plane of the converter is mounted onto a vertical surface, such as the back of an enclosure.
- *Recessed mounting*: The heat sinks on the back of the converter, projects through the back of the enclosure, into a cooling duct, and this allows the heat to be more effectively dissipated.

A sufficient separation from other equipments is necessary, to permit the unrestricted flow of cooling air through the heat sinks and across the electronic control cards. A general rule of thumb is that a free space of 100 mm should be allowed around all sides of the VSD. When more than one VSD is located in the same enclosure, they should preferably be mounted side by side rather than one above the other. Care should also be taken to avoid locating temperature-sensitive equipment, such as thermal overloads, immediately above the cooling air path of the VSD.

Adequate provision must be made to dissipate the converter losses into the external environment. The temperature rise inside the enclosure must be kept below the maximum-rated temperature of the converter.

General safety recommendations

The manufacturer's recommendations for installation should be carefully followed and implemented. The voltages present in power supply cables motor cables, and other power terminations are capable of causing a severe electrical shock.

In particular, the local requirements for safety, which are usually outlined in the wiring rules and other codes of practice, should always take priority over the manufacturer's recommendations. The recommended safety earthing connections should always be carefully installed, before any power is connected to the VSD equipment. AC VSDs have large capacitors connected across the DC link, as described in Chapter 3. After a VSD is switched off, a period of several minutes must be allowed to elapse before any work commences on the equipment. This is necessary to allow these internal capacitors to discharge fully. Most modern converters include some form of visual indication when the capacitors are charged.

Hazardous areas

In general, power electronic converters should not be mounted in 'hazardous areas', even when connected to an ex-rated motor, as this may invalidate the certification. When necessary, converters may be mounted in an approved enclosure and certification should be obtained for the entire VSD system, including both the converter and the motor.

6.9   Power supply connections and earthing requirements

6.9.1   Cable connection requirements

In accordance with accepted practice, power is normally provided to a VSD from a DB or a MCC. Adequate arrangements should be made to provide safety isolation switches and short-circuit protection at the power supply connection point. The short-circuit protection is required to protect the power cable to the AC converter and the input rectifier bridge at the converter. The converter provides the down-stream protection for the motor cable and the motor itself.

Adequate safety earthing should also be provided in accordance with the local 'Wiring Rules and Codes of Practice'. The metal frames of the AC converter and the AC motor should be earthed as shown in Figure 6.37 to keep touch potentials within safe limits. The chassis of the AC converter is equipped with one or more protective earth (PE) terminals, which should be connected to the common safety earth bar.

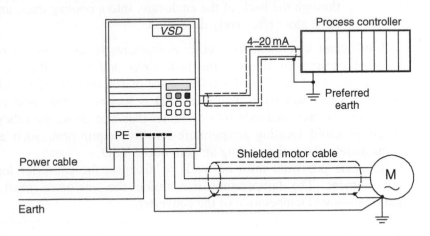

Figure 6.37
Power supply, motor and earthing connections

6.9.2   Power supply cables

The VSD should be connected to the power supply by means of a cable that is adequate for the current rating of the VSD.

Reference can be made to Australian Standard AS 3008 when selecting cables. The AC converter requires a three-phase supply cable (red/white/blue) and a PE conductor (green/yellow), which means a four-core cable with copper or aluminum conductors. A neutral conductor is unnecessary and is usually not connected to the frequency converter. The AC converter is a source of harmonic currents that flow back into the low impedance of the power supply system. This conducted harmonic current is carried into other electrical equipments, where it causes additional heat losses and interference.

Sensitive electronic instrumentation, such as magnetic flow-meters, thermocouples, and other microprocessor-based equipments, ideally should not be connected to the same power source, unless via a filtered power supply. In addition, interference can be radiated from the power supply cable and coupled into other circuits. These cables should, therefore, be routed well away from sensitive control circuits. The power supply cable should preferably be laid in a metal duct or a cable ladder and shielded in some way to reduce the radiation of emf due to the harmonic currents.

For this purpose, steel wire armored (SWA) cables are particularly suitable. If the power cable is unshielded, the control and communications cables should not be located within 300 mm of the power cable. The conductor sizes should be selected in accordance with the normal economic cable selection criteria. These usually take into account the maximum continuous current rating of the VSD, the short circuit rating, the length of the cable, and the voltage of the power supply system. The relevant local safety regulations should be strictly observed. However, when selecting the cable cross-sectional area for the power supply cables and upstream transformers, a de-rating factor of at least 10%

should be included to accommodate the additional heating due to the conducted harmonic currents.

If a supply side harmonic filter is fitted at the converter, this may not be necessary. Three-phase systems composed of three single-conductor cables should be avoided if possible. Power cables with a trefoil configuration produce a lower radiated emf.

6.9.3 Cables between converter and motor

The cable from the AC converter to the motor carries a switched PWM voltage, which is modulated at a high frequency by the inverter. This results in a higher level of harmonics than the power supply cable. Harmonic frequencies are in the frequency spectrum of 100 kHz–1 MHz. The motor cable should preferably be screened or located inside a metal duct. The control and communications cables should not be located close to this cable. The level of radiated emf is higher for cables, with three separate single cores, laid horizontally on a cable ladder, than a trefoil cable with a concentric shield.

The recommended size for the cable between the AC converter and the motor should preferably be the same as the power supply cable. The reasons are:

- It will be easier to add a bypass device in parallel with the frequency converter later, using the same cable, cable lugs, and connections.
- The load-carrying capacity of the motor cable is also reduced by harmonic currents and additionally by the capacitive leakage currents.

It should be borne in mind that the AC converter VSD provides short-circuit and overload protection, for the cable and the motor. A separate earth conductor between the converter and the motor is recommended for both safety and noise attenuation. The earth conductor from the motor should be reconnected to the PE terminal of the converter and not back to the DB. This will avoid any circulating high-frequency currents in the earth system.

When armored or shielded cables are used between the converter and the motor, it may be necessary to fit a barrier termination gland, at the motor end, when the cable is longer than 50 m. This is necessitated, because the high-frequency leakage currents flow, from the cable, through the shunt capacitance and into the shield. If these currents return via the motor or other parts of the earthing system, the interference is spread over a larger area. It is preferable, for the leakage currents to return to the source, via the shortest route, which is via the shield itself. The shield or SWA should be earthed at both the converter end and also at the frame of the motor.

6.9.4 Control cables

The control cables should be in accordance with the 'normal local practice'. These should have a cross-sectional area of at least 0.5 mm^2 for reasonable volt drop performance. The control and the communications cables, connected to the converter, should be shielded, to provide protection from the EMI. The shields should be earthed at one end only, at a point remote from the converter. Earthing the shield to the PE terminal of the drive should be avoided because the converter is a large source of interference. The shield should preferably be earthed at the equipment end.

Screened cables provide the best protection from coupled interference. The control cables should preferably be installed, on separate cable ladders or ducts and as far away from the power cables, as possible. If control cables are installed on the same cable ladder as the power cables, the separation should be as far as possible, with the minimum distance being about 300 mm. Long parallel runs on the same cable ladder should be avoided.

6.9.5 Earthing requirements

As mentioned earlier, both the AC converter and the motor must be provided with a safety earth, according to the requirements of local standards. The main purpose of this earthing is to avoid dangerous voltages, on the exposed metal parts, under fault conditions.

When designing and installing these earth connections, the requirements for the reduction of EMI should also be achieved, with these same earth connections. The main earthing connections of an AC converter are usually arranged as shown in Figure 6.37.

The PE terminal, on the converter, should be connected back to the system earth bar, usually located in the DB. This connection should provide a low impedance path back to earth.

6.9.6 Common cabling errors

The following are some of the common cabling errors made when installing VSDs:

- The earth conductor, from the AC converter, is run in the same duct or cable ladder, as other cables, such as the control and power cables for other equipments. Harmonic currents can be coupled into sensitive circuits. Ideally, instrument cables should be run in separate metal ducts or steel conduits.
- Running unshielded motor cables, next to the supply cable, to the AC converter or the power cables for other equipments. High-frequency harmonic currents can be coupled into the power cable, which can then be conducted to other sensitive electronic equipment. Other cables should be separated from the motor cable or converter power cable by a minimum of 300 mm.
- Cables between the AC converter and the motor should be no longer than 100 m. In case longer cables are used, motor filters are necessary to reduce the leakage current. Alternatively, the switching frequency may be reduced.

6.10 Precautions for start/stop control of AC drives

The protection requirements, for AC VSDs, have already been covered in previous chapters.

The protection of the mains supply side of the converter requires a short-circuit protection. This can be, either in the form of a set of adequately rated fuses, usually as part of a switch-fuse unit, or a main circuit breaker. The stop/start control of the AC drive can be achieved in a number of ways, mainly:

- Controlling the start/stop input of the converter control circuit
- Breaking the power circuit by means of a contactor.

The first is the recommended method, of controlling the stop/start of an AC converter. This may be achieved, by the stop and start pushbuttons being wired directly to the control terminals of the converter, as shown in Figure 6.38. Alternatively, if the control is from a remote device such as a PLC, it can be wired from the PLC directly to the terminals of the AC converter, as shown in Figure 6.39. The second method is, the one most commonly used, for the DOL starting of normal fixed-speed AC motors. Following from the previous DOL 'standard' practice, this method is also quite commonly used in the industry for the control of VSDs, particularly for conveyors.

It is usually a safety requirement, to provide an emergency stop or pull-wire switch, to interrupt the power circuit when operated. While this method satisfies the safety requirements by breaking the power supply to the motor, there are a number of potential hazards with this method of control.

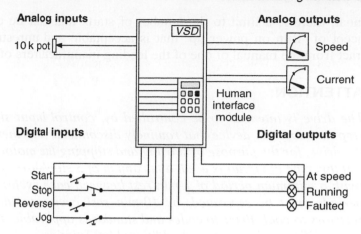

Figure 6.38
Configuration of a typical hard-wired manual control system

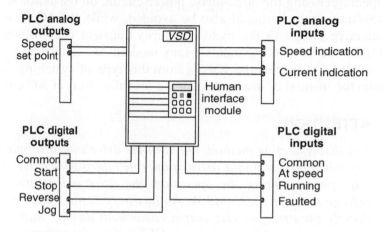

Figure 6.39
Configuration of a typical hard-wired automated control system

The main problems are:

1. *Contactor on supply side of the AC converter*

Opening/closing the supply side of the AC converter for stop/start control should be avoided because most modern converters take their power from the DC Bus. Every time the power is removed, the following takes place:

- Power to the control circuit is lost
- Control display goes off
- Diagnostic information disappears
- DC capacitors become discharged
- Serial communication is lost.

When the AC VSD needs to be restarted, there is a time delay (typically 2 s), while the DC Bus charging system completes its sequence to recharge the DC capacitor. This stresses the charging resistors, the DC capacitor, and other components. The charging resistors of many AC converters are short time rated. Although not highlighted in the user manual, there is a limit to the number of starts that can be done. Many users have the concept of 'Run on power up' that is acceptable and unrestricted. The following is an extract from the manual of one of the leading manufacturers of AC converters:

ATTENTION:

The drive is intended to be controlled by, control input signals that will start and stop the motor. A device that routinely disconnects and then reapplies line power to the drive, for the purpose of starting and stopping the motor, is not recommended.

If this type of circuit is used, a maximum of 3 stop/start cycles, in any 5-min period (with a minimum period of 1-min rest between each cycle) is required. These 5-min periods must be separated, by 10-min rest cycles, to allow the drive precharge resistors to cool. Refer to codes and standards applicable, to your particular system, for specific requirements and additional information.

2. *Contactor on motor side of the AC converter*

Opening/closing the three-phase power circuit, on the motor side of the AC converter, for stop/start control, should also be avoided, while the AC drive is running. Breaking the inductive circuit to the motor produces transient over-voltages, which can damage the IGBTs and other components. Many modern AC converters have RC suppression circuits (snubbers) to protect the IGBTs from this type of switching. The following is an extract from the manual of one of the leading manufacturers of AC converters:

ATTENTION:

Any disconnecting method, wired to the drive output terminals U, V and W must be capable of disabling the drive, if operated during the drive operation.

If opened, during the operation, the drive will continue to produce an output voltage between U, V and W. An auxiliary contact must be used, to simultaneously disable the drive, or else output component damage may occur. The objective is to ensure, that the AC Converter is OFF before the contacts between the converter and the motor are opened. This will avoid IGBT damage due to transient over-voltages. In addition, closing the motor side contactor, when converter output voltage is present, can result in a motor inrush current, similar to DOL starting. Apart from the stress this places on the converter, the drive will trip on over-current. Repeated attempts at closing the motor contactor after the converter has started, may eventually lead to IGBT failure.

If a contactor has to be installed into the power circuit of an AC Variable Speed Drive system, to meet local safety requirements, then it is better to locate this contactor downstream of the AC converter. It is then necessary to include, an auxiliary contact on the contactor, which disables the converter control circuit, BEFORE the contactor is opened or closes the enable circuit, AFTER the contactor has been closed. This means that a Late Make-Early Break auxiliary contact, should be used on the contactor and wired to the converter Enable input.

While the above configuration will protect the AC converter from failure, this method of routine stop/start control is NOT recommended. It should be used for "Emergency Stop" conditions only. Routine stop/start sequences should be done from the AC converter control terminals. An alternative method of ensuring that plant operators follow this requirement is, to install a latching relay and a Reset pushbutton. The latching relay needs to be reset after every 'Emergency Stop' sequence.

6.11 Control wiring for VSDS

VSD may be controlled 'locally' by means of manual pushbuttons, switches, and potentiometers mounted on the front of the converter. For simple, manually controlled operations, these local controls are all that is required to operate the VSD.

In most industrial applications, it is not practical to control the VSD from the position where the VSD is located. VSDs are usually installed inside MCCs, which are located in switch rooms, usually close to the power supply transformer, but not necessarily, close to where the operator is controlling the process.

6.11.1 Remote operation of VSD

Almost all VSDs have terminals that permit 'Remote Control' from a location close to the operator. VSDs have terminals for the following controls:

- *DI*, such as remote Start, Stop, Reverse, Jog, etc., which are usually implemented by

  (a) Remote pushbuttons in a manually controlled system
  (b) DO of a process controller in an automated system.

- *Digital status outputs*, such as indication of Running, Stopped, At Speed, Faulted, etc., which are usually implemented by

  (a) Remote alarm and indication lamps in a manually controlled system
  (b) DI to a process controller in an automated system.

- *AI*, such as remote Speed Reference, Torque Reference, etc., which are usually implemented by

  (a) Remote potentiometer (10 kΩ pot) in a manually controlled system
  (b) Analog outputs, such as remote Speed Indication, Current Indication, etc., which are usually implemented by remote display meters (0–10 V) in a manually controlled system
  (c) AI to a process controller in an automated system, usually using a 4–20 mA signal carried on a screened twisted pair cable.

Manual and automated control systems have operated very effectively for many years with this type of 'hard-wired' control system. The main disadvantage of this system is:

- All DIs and DOs require one wire per function and a common connection
- All AIs and AOs require two wires per function and a shield connection.

6.11.2 Hard-wired connections to PLC control systems

With the introduction of automated control systems using, PLCs and distributed control systems (DCS), the 'hard-wired' control connections have been extended, with input/output (I/O) modules replacing the manual controls.

Control systems have grown in complexity and the amount of information required from field sensors has expanded the number of conductors required to implement the automated control system. This affects both the cost and the complexity.

As more field devices become integrated into the overall control system, there arises the problem of complex cabling.

A hard-wired interface between a VSD and a PLC would typically require about 15 conductors as follows:

- Five conductors for controls such as Start, Stop, Enable, Reverse, etc.
- Four conductors for status/alarms, such as Running, Faulted, at Speed, etc.
- Two or three conductors for analog control, such as Speed set point
- Four conductors for analog status, such as Speed indication, Current indication.

If there are several VSDs in the overall system, the number of wires is multiplied by the number of VSDs in the system.

6.11.3   Serial communications with PLC control systems

'Serial Communications' helps overcome these problems and allows complex field instruments and VSD systems to be linked together, into an overall automated control system with minimum cabling.

Microprocessors-based digital control devices, sometimes called 'Smart' devices, are increasingly being used in modern factory automation and industrial process control systems. Several 'smart' devices can be 'multi-dropped' or 'daisy-chained' on one pair of wires and integrated into the overall automated control system. Control and status information can be transferred serially between the process controller and the VSDs located in the field. Parameter settings can also be adjusted remotely from a central point (Figure 6.40).

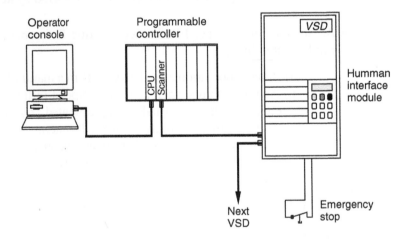

Figure 6.40
Configuration of a typical serial communications system

6.12   Commissioning VSDs

6.12.1   The purpose of commissioning

The main purpose of commissioning VSDs is to ensure the following:

- The AC converter and motor have been correctly installed and meet the wiring and safety standards such as AS 3000.
- The power and motor cables are correctly sized, installed, and terminated.

- All power cable shields have been correctly earthed at both ends to the PE terminal at the converter, at the motor, and at the DB or MCC.
- The control cables have been installed according to the control system design.
- All control cable shields have been correctly earthed at one end only, preferably at the process control system end ('cleaner' earth).
- There are no faults on the cables prior to energizing for the first time.

6.12.2 Correct application settings selection

Once all the basic checks have been completed, along with the commissioning test sheet, the VSD is ready for energizing.

It is recommended that when energizing the converter for the first time, the motor cables should be disconnected until all the basic parameter settings have been installed into the converter. This will avoid problems like starting the motor in the wrong direction, starting with a high acceleration time, etc. There is no danger in running a PWM converter with the output side in a completely open circuit. Once all the initial settings and on-load checks have been completed, the motor cable is insulation-tested and connected for the final on-load commissioning tests.

6.12.3 Correct parameter settings selection

A VSD will only perform correctly, if the basic parameters have been correctly set, to suit the particular application. The following are the basic parameters that must be checked, before the VSD is connected to a mechanical load:

- The correct base voltage must be selected for the supply voltage and to suit the electric motor connected to the output. In Australia, this standard voltage is usually 415 V, three phase. This will ensure that the correct output Volts/Hz ratio is presented to the motor.
- The correct base frequency must be selected for the supply voltage and to suit the electric motor connected to the output. In Australia, this standard frequency is usually 50 Hz. This will ensure that the correct output Volts/Hz ratio is presented to the motor.
- The connections to the cooling fan should be checked to ensure that the correct tap on the transformer has been selected.

Thereafter, the remaining parameter settings can be selected as follows:

- *Maximum speed*:   Usually set to 50 Hz, but often set to a higher speed to suit the application. Reference should be made to Chapter 6 to ensure that the maximum speed does not take the drive beyond the loadability limit.
- *Minimum speed*:   Usually 0 Hz for a pump or fan drive, but often set at a higher speed to suit constant torque applications. Reference should be made to Chapter 6 to ensure that the minimum speed does not take the drive below the loadability limit.
- *Rated current of the motor*:   This depends on the size of the motor relative to the rating of the converter. The current rating of the converter should always be equal to, or higher than, the motor rating. For adequate protection of the motor, the correct current rating should be chosen.
- *Current limit*: Determines the starting torque of the motor. If a high breakaway torque is expected, a setting of up to 150% will provide the highest starting torque.

- *Acceleration time*: Determines the ramp-up time from zero to maximum speed. This should be chosen in relation to the inertia of the mechanical load and the type of application. For example, in a pumping application, the acceleration time should be slow enough to prevent water hammer in the pipes.
- *Deceleration time*: Determines the ramp-down time from maximum speed to zero. This setting is only applicable if the 'ramp to stop' option is selected. Other alternatives are usually 'coast to stop' and 'DC braking'. On high inertia loads, this should not be set too short. If the deceleration time is below the natural rundown time of the load, the DC voltage will rise to a high level and could result in an unexpected tripping on 'over-voltage'. The deceleration time can only be shorter than the natural rundown time if a dynamic braking resistor has been fitted.
- *Starting torque boost*:   Can be selected if the load exhibits a high breakaway torque. This feature should be used cautiously, to prevent overfluxing of the motor at low speeds. Too high a setting can result in motor overheating. Only sufficient torque boost should be selected, to ensure that the VSD exceeds the breakaway torque of the load during starting.

There are many settings commonly required on modern digital VSDs. The above are the most important and must be checked before starting. The remaining parameters usually have a 'default' setting which will probably be adequate for most applications. However, these should be checked and adjusted for optimum operation.

This chapter has detailed installation and commissioning guidelines for VSD for AC motors. If these guidelines are followed in all aspects, then it will definitely help during installation, commissioning, and troubleshooting of VSDs.

7

Troubleshooting control circuits

Objectives

- To understand the fundamentals of control circuits
- To understand types of control circuits
- To troubleshoot control circuits.

7.1   Basic control circuits

Basic control circuits are used in starting, stopping, sequencing, and safety automatic interlocking of equipment and machines.

The control circuit consists of relays, relay contacts, contactors, timers, counters, etc. Control circuits can also be configured or programed in the PLCs. This is done, using ladder logic diagram, statement lists, or control flowcharts software, by representing the logical conditions, sequences, and interlocks required for operating equipment or a machine in an automatic sequence.

To understand how to troubleshoot control circuits, it is very important to understand the working of some basic control circuits, as given in Figure 7.1.

7.1.1   Basic control circuit for DOL (direct-on-line) starter

Figure 7.1(a) shows a typical circuit for a DOL starter for a three-phase motor. A full-line voltage is applied across the windings with this starter. The rating of motors which can be started direct-on-line depends on the capacity of the distribution system and the acceptable bus voltage drop during starting. In large industrial systems it is not unusual for even a 200 kW motor to be DOL started especially when fed by a transformer of 1600 kVA or higher. However, when a motor is fed by an LV emergency generator, DOL starting has to be planned with due consideration to starting voltage drop.

Main circuit

Figure 7.1(a) shows the circuit with a three-phase power supply (L1, L2, and L3), main circuit fuses (F1), main contactor (K1), and an overload protection relay (F2) for a three-phase motor.

The motor can be started by the following two methods:

1. Momentary contact control with press and release type pushbutton
2. Maintained contact control with press and latch type pushbutton.

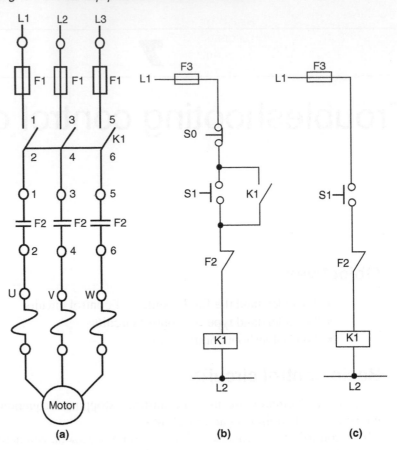

Figure 7.1
(a) *Main circuit;* (b) *Control circuit (Momentary contact);* (c) *Control circuit (Maintained contact)*

Momentary contactor control

Figure 7.1(b) shows a momentary control circuit to start and stop the three-phase motor using a DOL starter with a start and stop pushbutton S1 and S0 respectively.

The control circuit consists of an overload relay (F2) NC contact, an NC contact of stop pushbutton (S0), NO contact of start pushbutton (S1) connected in series to the main contactor (K1) coil. The control supply for the circuit passes through a control fuse (F3).

The main contactor coil gets the phase line (*L1*) through the control circuit only when all the contacts are closed. In this case, when the start pushbutton is pressed, the control circuit is closed and the main contactor is energized. As shown in Figure 7.2, an NO contact of the main contactor is connected in parallel to the start pushbutton. As the main contactor is switched on, it is latched through this parallel NO contact (K1) even after the start pushbutton is released. The main contactor remains on and the motor continues to run until the stop pushbutton is pressed to stop the motor or the motor trips due to an overload relay operation.

Maintained contactor control

Figure 7.1(c) shows a control circuit to start and stop the three-phase motor using a DOL starter with a single pushbutton (S1).

The control circuit consists of an NC contact of overload relay (F2) and NO contact of a toggle type switch (S), connected in series to the main contactor (K) coil. The control supply for the circuit passes through the control fuse (F3).

The main contactor (K1) coil gets power only when all the contacts are closed. In this case, when the switch (S) is closed, the control circuit closes and the main contactor (K1) is energized.

As long as switch (S) is maintained on, the main contactor remains on and the motor continues to run until the switch (S) is opened or the motor trips due to an overload relay operation.

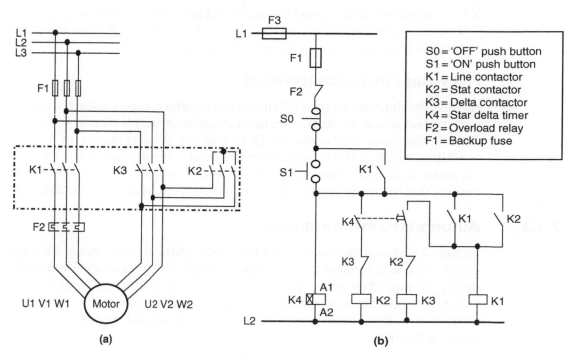

Figure 7.2
(a) Main circuit of star–delta starter; (b) Control circuit of star–delta starter

7.1.2   Star–delta 3ϕ starter

The circuit shown in Figure 7.2(a) is the main circuit for the star–delta starter and Figure 7.2(b) is the control circuit.

Usually, a motor has the tendency to draw 500% higher current than the full load of normal current from the supply line during startup. This in turn increases the starting torque that is higher than normal, which can result in a mechanical damage. To avoid this, reduced voltage starters are used. Star–delta starters are also used when a weak system cannot support the DOL starting of a large capacity motor. The starting current (line) when using this method is reduced by factor of 3 (i.e. 200% in place of 600%). The starting torque however also reduces by a factor 3. This method is therefore not suitable for loads with high inertia or those that require high starting/break away torque.

During the startup in a star–delta starter, the winding is connected in a star configuration with contactor K1 and K2, which applies reduced voltage (approximately 58% of rated). Then after a while, connect the windings in delta configuration with contactor K1 and K3.

Star–delta starter working

The main contactor K1 will energize only when the control circuit fuse (F3), backup fuse (F1), and the overload relay (F2) are healthy and the start pushbutton (S1) is pressed.

Reduced-voltage configuration (star configuration)

Star–delta timer coil (K4) gets power through fuses F3, F1, NC contact of stop pushbutton (S0), and NO contact of start push button. As start PB (S1) is pressed, the timer coil K4 will pickup and in turn energize the star contactor coil K2. The main line contactor (K1) coil gets power via the NC contact of S0, NO contact of S1, NO contact of K2 and remains latched unless the stop pushbutton (S0) is pressed.

Now, the main line contactor (K1) and the star contactor (K2) are in a pickup condition, which will drive the motor in the star configuration.

Full voltage (delta configuration)

As the time duration set on a K4 timer (star to delta timer) elapses, the contactor coil (K3) is picked up and at the same time, the star contactor (K2) is de-energized.

Now, the main line contactor (K1) and the delta contactor (K3) are in a pickup condition, which will drive the motor in a delta configuration. When the motor trips in an overload condition either in a star or delta configuration, the control circuit always ensures that the motor restarts in a star configuration, rather than the delta configuration.

7.1.3 Autotransformer 3 ϕ starter

Figure 7.3 shows the autotransformer three-phase starter circuit. This type of starter circuit uses an autotransformer to apply reduced voltage across the windings of the motor during startup. Three autotransformers are connected in the star configuration and taps are selected, to provide an adequate starting current for the motor.

After a certain time lapse, full voltage is applied to the motor bypassing the autotransformers.

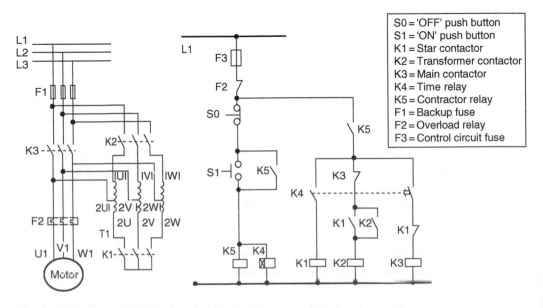

Figure 7.3
Typical main and control circuit of an autotransformer starter for a three-phase motor

The working of an autotransformer

The fuse (F1) and the overload relay (F2) provide protection to the main circuit. Similarly, the control circuit has the fuse (F3) and overload relay (F2) NC contact.

Reduced voltage configuration

In this circuit, the contactor (K5) will pickup when the start push button is pressed and will remain latched until the stop push button is pressed or the control circuit fuses or the motor trips on overload.

As K5 picks up, it will energize the timer relay (K4) coil. This in turn will energize the contactor K1 coil. The closing contactor K1 contact will energize the contactor K2 coil. So, contactors K5, K4, K1, and K2 are in an energized condition at this stage. This will result in starting the motor through an autotransformer at a reduced voltage and with star configuration because of the contactor K1 and K2.

Full voltage configuration

As timer relay (K4) time lapses, it de-energizes the contactor K1 coil. At the same time contactor K3 coil is energized, this will in turn de-energize the contactor K2 coil.

The motor will now run at full voltage as contactor K3 is in a pickup condition. If in the interim, the motor trips on overload, then the control circuit has to be checked, so the motor restarts in a star configuration and at a reduced voltage after the overload reset.

7.2   Ladder logic circuits

The designing, programing, testing, commissioning, troubleshooting, and maintenance of control logic are much easier using the ladder logic programs in a PLC than in hard-wired circuit.

We have discussed the ladder logic instructions before and looked at the simple ladder logic programs in Chapter 2. Let us consider a ladder logic program for a typical control circuit of the DOL starter for a three-phase motor as shown in Figure 7.1(a).

For a three-phase motor with a DOL starter, the following input and output signals are configured in the PLC.

7.2.1   Digital inputs

1. Control voltage ON and control fuse 'F3' OK (normally open)
2. Motor overload 'F2' (normally closed)
3. Motor stop 'S0' (normally closed)
4. Motor start 'S1' (normally open)
5. Main contact on feedback (normally open).

7.2.2   Digital output

Main contactor on

The ladder logic program Instructions for the DOL starter control circuit with a maintained contact control is shown in the Figure 7.4.

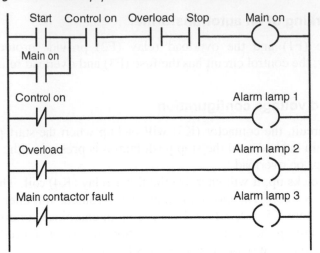

Remarks:

1. 'Start' push button contact is closed as push button is pressed momentarily.
2. 'Control on' input will be on, if line voltage is there and control fuse is not blown.
3. Overload relay NC contact will keep the input 'Overload' on till the relay is not operated.
4. 'Stop' input from stop push button NC contact will remain on till stop push button is not pressed.
5. Contact of 'Main on' will hold the coil output 'Main on' until the stop push button is pressed or overload relay trips.

Figure 7.4
Ladder logic program for a DOL starter and alarms

As various inputs for the control circuit in the PLC has been gathered, it will be for generating alarms. These alarms can be indicated with the help of 'Indication Lamps' that can be mounted on the motor starter panel in addition to the start and stop pushbuttons. These alarms can help in faultfinding. The following alarms can be configured using the inputs already available in the PLC. The PLC outputs can be generated to drive these alarm indication lamps.

In case of the DOL starter for a three-phase motor, the following alarms can be configured to indicate faults:

- *Alarm indication lamp-1*:   Control voltage OFF
- *Alarm indication lamp-2*:   Motor tripped on overload
- *Alarm indication lamp-3*:   Main contactor feedback fault.

7.3   Two-wire control

The circuit shown in Figure 7.5 can be used for an auto start operation of a motor after a power failure depending on the position of the control contact. The control contact may be a level switch contact or a temperature switch contact.

The contactor K remains energized as long as the control contact is in a close condition. If the power failure occurs, the motor turns off.

The motor will then restart automatically, once the power is restored since the control contact is in a close condition.

This circuit type is useful for pumps, fans, blowers, etc. where an automatic restarting of the device after a power failure is desirable depending on the control contact.

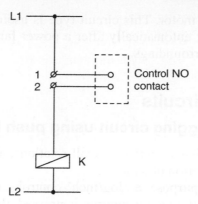

Figure 7.5
Two-wire control circuit

7.4   Three-wire control – start/stop

A three-wire control circuit for start/stop operation is shown in Figure 7.6. As shown in the diagram, NO contact of the start pushbutton and NC contact of the stop pushbutton are in series with the main contactor in the control circuit.

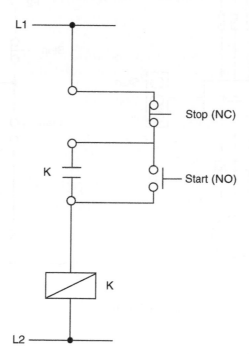

Figure 7.6
Three-wire control – start/stop

The main contactor K will energize as a start pushbutton is pressed and if the stop pushbutton is not pressed, it stays latched. The latching contact K is used for the main contactor.

If a power failure occurs and the motor stops, then the motor will not start automatically. Again, the start pushbutton has to be pressed to latch the main contactor and to start the motor. This circuit type is useful in applications where the motor is not required to start automatically after a power failure so as to prevent the occurrence of a hazard to the surroundings.

7.5   Jog/inch circuits

7.5.1   Start/stop/jogging circuit using push buttons

If the machine has to perform small rotations, a control circuit is required to accomplish the 'Inching' motion of a motor.

To serve the purpose, a 'Jog/Inch' control circuit for a three-phase motor is designed such that when the jog pushbutton is pressed, the motor runs and when the pushbutton is released the motor stops (Figure 7.7). Generally, this type of movement is used in machine tools.

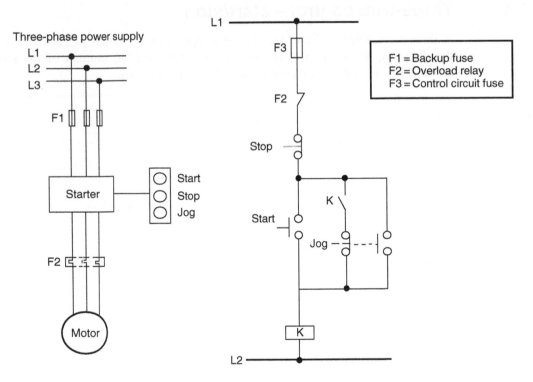

Figure 7.7
Start/stop/jogging circuit using pushbutton

Operation

As shown in the control circuit, a double-contact jog pushbutton is used with one NC contact and one NO contact. Therefore, when the jog button is pressed, the latching circuit to the starter coil (K) is opened by the NC contacts of the jog pushbutton. Therefore, the starter coil (K) will not lock in; rather it remains energized as long as the jog button is fully pressed. Thus, an 'Inching/Jogging' action can be achieved.

If the jog pushbutton were released suddenly, then if the NC contacts closes before the starter maintaining the contacts (K) open, the motor would continue to run. This in turn can prove hazardous to the surrounding workers and machinery. A mechanical device can be installed which ensures that the starter-holding circuit is not reestablished if the jog button is released too rapidly.

7.5.2 Start/stop/jogging circuit using selector switch

Figure 7.8 shows the use of a selector switch in the control circuit to obtain jogging. The start button performs twin functions; it works as a jog button as well as a start button.

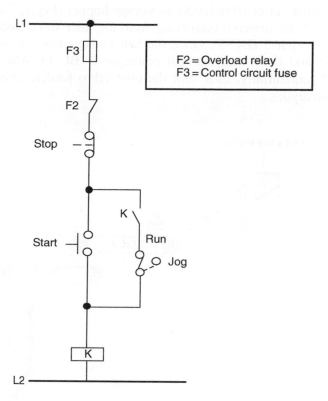

Figure 7.8
Start/stop/jogging circuit using selector switch

To operate the motor in a run mode, the selector switch should be in the 'RUN' position. The K coil circuit is completed if the start button is pressed and it remains latched because of the latching contact K1 and the selector switch.

To operate the motor in the jog mode, the selector switch should be in the jog position. If the start button is pressed, the K coil circuit is completed but as soon as the start button is released, the K coil is de-energized because the latching circuit of K1 coil is open.

7.6 Sequence start and stop

In large industrial plants, there are a large number of machines and drives. They are required to start and stop in a predetermined sequence. In such cases, it is impossible to start/stop each drive with an individual start/stop control circuit. Moreover, it is impossible to monitor each drive, follow the sequence of start/stop for the drives, and stop drives as per the process inputs.

To perform this kind of operation, sequencers are used; these may be mechanical or electromechanical.

With 'Time' sequencers, depending on the time duration and base sequencers, a number of outputs are switched on-off in a predefined sequence.

In the case of 'pulse' sequencers, a number of outputs are switched on-off in predefined sequences, depending on the pulses received from the process (pulses may be either derived from the 'proximity' switches or the 'limit' switches, etc.).

To understand this, consider a sequence of the three-belt conveyor system carrying raw coal from an inlet vibro feeder to storage hopper (Figure 7.9).

To start the material conveying from the inlet vibro feeder to the hopper, first start the belt conveyor-3 (BC-3). Once the belt conveyor-3 is running, start the belt conveyor-2 (BC-2) and then start the belt conveyor-1 (BC-1). After the three conveyer belts are started sequentially, then start the inlet vibro feeder, which drops the material onto the belt conveyors.

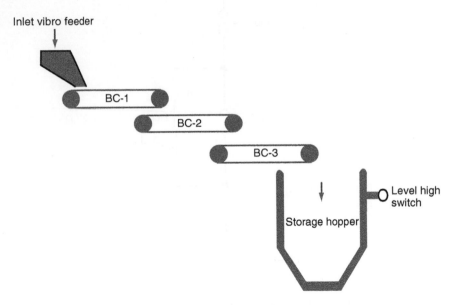

Figure 7.9
Sequence of belt conveyors

During the material feeding, if BC-3 or BC-2 trips, the upstream belt conveyors must also be stopped immediately. In addition, when the raw material hopper is filled to a high level, the belt conveyors must stop in the reverse sequence.

The sequence for start and stop for the 'Belt Conveyor System' is shown in the Figure 7.10.

Note: The purpose of not stopping conveying equipment immediately on receiving a STOP command is to avoid material remaining on the conveying equipment. Therefore in some installations, the material feed (such as a gate) is closed immediately on receipt of a STOP command and then the whole line is stopped after a preset delay adequate for the entire material remaining in the system to be cleared. This scheme avoids the necessity of multiple timer circuits required for a sequential stopping.

Start sequence   **Stop sequence**

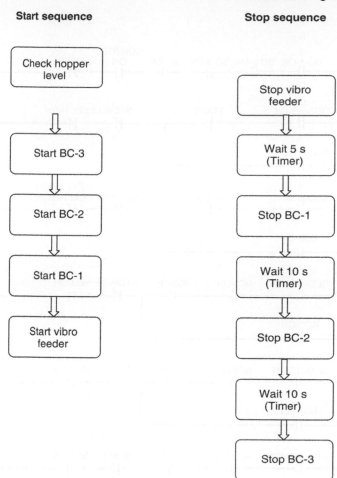

Figure 7.10
Start and stop sequence for the belt conveyors system

7.7   Automatic sequence starting

It is difficult to manually control (start/stop) multiple drives simultaneously, while depending upon the various interlocks. Sometimes a large number of machine drives, in a plant or section of a plant, are required to run in a predefined sequence. When these are interlocked with each other, it can be quite difficult to start/stop the drives manually in sequence, depending on the interlocking that may be required for safety or due to the process devices.

A typical automatic belt conveyor sequence system is shown in Figure 7.11. An automatic sequence start and stop of the belt conveyors is done by using a single start push button, a stop push button, a high level switch in the hopper, and timers for auto sequencing.

7.8   Reversing circuit

An interchange of any two phases reverts the direction of a three-phase motor. This will run the motor in the reverse direction.

To accomplish this reversal, two different types of control circuits are detailed as follows.

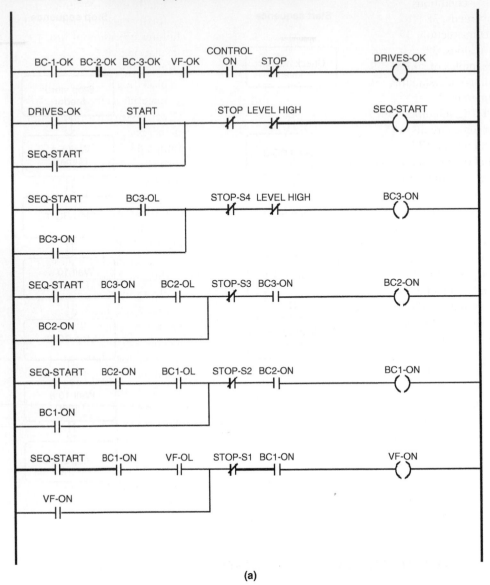

(a)

Figure 7.11
Automatic sequence starting and stopping: (a) Start sequence; (b) Stop sequence

7.8.1 Jog type for/rev/off circuit using selector switch

Interchanging any two leads to a three-phase induction motor will cause it to run in the reverse direction. A three-phase reversing starter shown in the main circuit (Figure.7.12) shows two contactors K1 and K2 (forward and reverse, respectively). The selector switch is of the spring return-type with the center off.

The contactor coil K1 is energized keeping the selector switch in a forward position. The contactor K1 connects the supply leads (L1, L2, and L3) to the motor leads (U, V, and W) in the same phase sequence. This causes the motor to run in a forward direction. Keeping the selector switch in a reverse position energizes the contactor coil K2. The contactor K2 connects the supply leads L1 to W, L2 to V, and L3 to U changing the phase sequence of L1 and L2. This causes the motor to run in a reverse direction. Putting the selector switch in an off position turns off the motor.

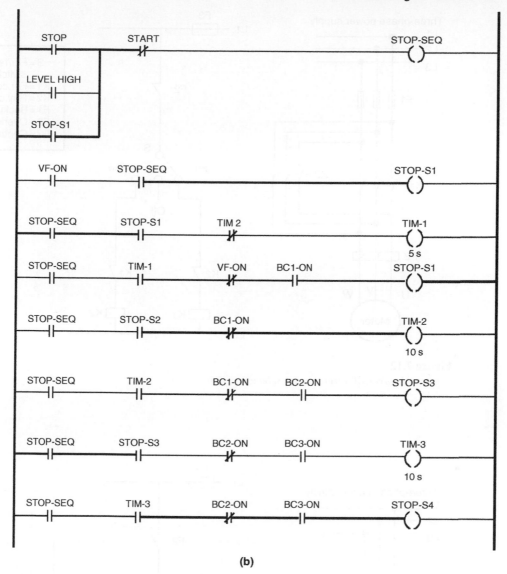

(b)

Figure 7.11 (continued)

The motor is protected from a short-circuit condition if an overload relay, a backup fuse, and a control circuit fuse provide both the forward and reverse protection to the motor.

7.8.2 Latch type for/rev/stop circuit using pushbuttons

The circuit discussed earlier was of the jog type, fwd/rev circuit. The circuit shown in Figure 7.13 is of the latch type, fwd/rev control.

Pressing the forward pushbutton will energize the contactor K1 coil. This in turn will connect the supply leads to the motor leads in the same phase sequence causing the motor to rotate in a forward direction.

The contactor K1 will remain energized because of the latching contact of K1. The motor will continue to run in a forward direction until the stop/reverse pushbutton is pressed or the motor trips on overload or the protection fuse blows.

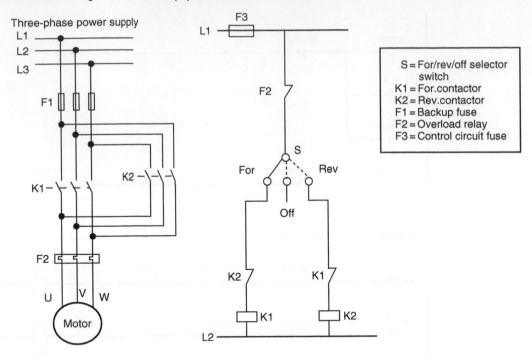

Figure 7.12
Jog type fwd/rev/off circuit using selector switch

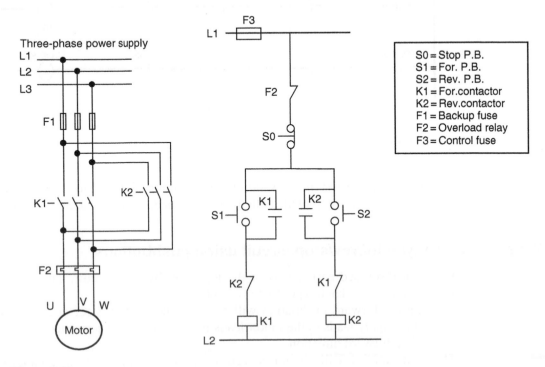

Figure 7.13
Latch-type fwd/rev/off circuit using push buttons

Pressing the reverse pushbutton will energize the contactor K2 coil simultaneously by de-energizing the K1 coil. This will connect the supply leads to the motor leads in a different phase sequence causing the motor to rotate in a reverse direction. The contactor K2 will remain energized because of the latching contact of K2. The motor will continue to run in a reverse direction until the stop/forward pushbutton is pressed or the motor trips on overload or the protection fuse blows.

The stop button need not be pressed before changing the direction of the rotation.

7.9 Plug stop and anti-plug circuits

To halt a motor or to stop a running motor, a common method is to remove the supply voltage and allow the motor and load to come to a stop.

Nevertheless, in some applications, the motor must be stopped instantaneously or held in position by some sort of a braking device.

This is achieved using the electric braking circuit. It uses the windings of the motor to produce a retarding torque. The kinetic energy of the rotor and the load is dissipated as heat in the rotor bars of the motor.

The following are the two different means of electric braking:

1. Plugging
2. Dynamic braking.

7.9.1 Plugging

In this method, a motor is connected to run in a reverse direction. This is done while the motor is still running in a forward direction, resulting in motor stoppage.

In order to achieve this, a switch, or a contact is used which gives the status of the motor. Depending on the running of a motor and its speed, the switch status changes from NO to NC.

This switch is called the zero-speed switch or the plugging switch. A zero-speed switch prevents a motor from reversing after it has come to a stop.

A 'zero-speed' switch is physically coupled to a moving shaft on the machinery, the motor of which is to be plugged. As the zero-speed switch rotates along with the machine, the centrifugal force causes the contacts of the switch to open or close, depending on its intended use.

Each zero-speed switch has a rated operating speed range, within which the contacts will be switched, example, 10–100 rpm. The control schematic of Figure 7.14 shows one method of plugging a motor to stop from one direction only.

As the start (forward) pushbutton is pressed, it energizes forward the contactor coil K1. Therefore, the motor runs in the forward direction. The contactor K1 is latched through its latching contact.

As the motor runs in the forward direction, the NC contact F (zero switch) opens the circuit of the reverse contactor coil K2. If the stop pushbutton is pressed, it will de-energize the forward contactor K1. This in turn will help the reverse contactor K2 to energize because the forward contact on the speed switch is also in a closed condition.

As the reverse contactor is energized, the motor is plugged. The motor starts decreasing in speed rapidly up to the setting of the speed switch, at which point its contact opens and de-energizes the reverse contactor K2.

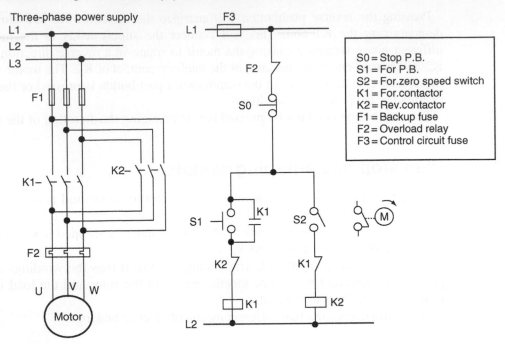

Figure 7.14
Plugging circuit for three-phase motor

This contactor is used, only to stop the motor, using the plugging operation. It is not used to run the motor in reverse. Many machines require the motor be able to reverse. Most small machines are not adversely affected by reversing the motor, before coming to a stop.

This is not true of larger pieces of equipment. The sudden reversing torque applied when a large motor is reversed (without slowing the motor speed) could damage the motor.

The driven machinery and the extremely high current could affect the distribution system. Plugging a motor for more than five times a minute requires the motor starter to be de-rated.

7.9.2   Anti-plugging

Anti-plugging protection is necessary, when a motor with large inertia is connected suddenly, in a reverse direction, while the motor is still running in a forward direction.

Anti-plugging protection prevents the application of a counter torque, until the motor speed is reduced to an acceptable value. In the anti-plugging circuit shown in Figure 7.15, the motor can be reversed but not plugged. Pressing the forward pushbutton completes the circuit for the K1 contactor coil causing the motor to run in the forward direction. It continues to run because of the latching contacts of K1. With the NC contact F (zero-speed switch contact) in reverse, the contactor K2 opens, because of the forward rotation of the motor.

Pressing the stop button de-energizes the K1 contactor coil, which opens the latching contact of K1 also, causing the motor to slow. Pressing the reverse button will not complete a circuit for the K2 contactor coil until the F (zero-speed switch) contact re-closes (i.e., when the speed is reduced below switch setting).

Therefore, only when the motor reaches a near-zero speed, can the reverse circuit be energized. The motor now runs in a reverse direction.

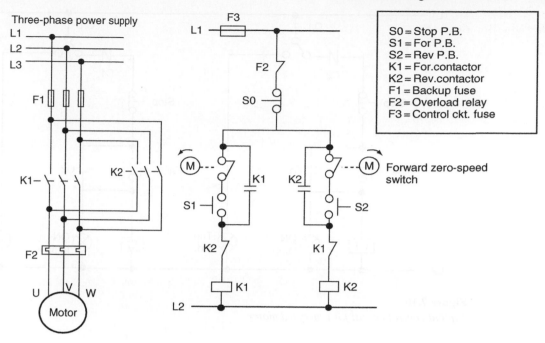

Figure 7.15
Anti-plugging circuit for three-phase motor

7.10   Two-speed motor control

At times, it is required to run equipment at two different speeds. This is usually so in certain industrial applications, such as mixer motor speeds, ventilating pumps, and batch control processes. Particularly in batch control, while feeding the batch components in a mixer, these components are fed into the mixer at two different feed rates – fast coarse feeding and slow fine feeding. This is done to feed the components accurately and to avoid overshoots. To achieve this, two-speed motors are employed.

A typical control circuit for a two-speed motor is shown in Figure 7.16. There are two electrically separate windings housed in the motor. The control circuit connects the windings in different configurations causing the speed to change from one rpm to another. Each winding can deliver the motor's horsepower at a rated speed.

As shown in Figure 7.16, two contactors are incorporated for low and high rpm. They should not be activated at the same time electrically. To protect both of them separately, an independent overload relay protection is provided.

7.11   Overload protection

Protection is provided to protect the motor from excessive overheating due to a motor overload. The overload protection protects the motor from excessive loading and at the same time protects starter components and conductors from overheating.

The working principle being, to sense the current flowing through the motor, that being a direct measure of heating and load to the motor. For overload protection in a motor circuit, bimetal relays are commonly used. The bimetal relays are triple-pole adjustable overload relays with built-in single-phasing protection. They provide accurate and reliable protection to the motors against overload and single phasing.

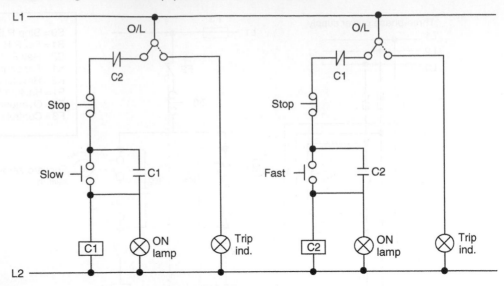

Figure 7.16
Typical control circuit for two-speed motor

The bimetal relays provide an accurate overload and accelerated single-phasing protection for the motors. They incorporate the duel slider principle for accelerated tripping under the single-phasing protection. They also provide protection against severe unbalanced voltages.

The thermal relay is connected in series with the motor conductors, as excessive current passes through it for a preset time interval, a contact (in series) gets operated resulting in a motor trippage. The relay has settings (generally 3–14% of full load current) for various current ratings of the motor. These can be set accordingly.

Following the curve of a thermal overload, the relay shows the relationship between current and time. The relay will not trip at a rated current, but at twice the rated current after a time duration of 45 s approximately. The bimetal relays protect themselves against overloads of up to 10 times the maximum setting. Beyond this limit, they have to be protected from short-circuits. It is mandatory to use backup fuses for this purpose. The typical operating characteristics of bimetal relays are shown in the Figure 7.17.

In high intertia drives, the starting time of the motor may be high (30–60 s). Use of standard thermal overload relays will cause the motor to trip before it achieves rated speed. In such cases, special thermal overload relays with a saturable CT may be required and are available from several vendors. A backup protection may be also provided in the form of a speed switch and a timer to detect abnormal starting condition and trip the motor.

7.12   Troubleshooting examples

In the above sections, various basic and complex control circuits for a three-phase motor have been dealt with.

The following is an example of troubleshooting a control circuit. Consider the control circuit for a three-phase motor with a DOL starter with a maintained-contact control as shown in Figure 7.1(b).

3-ϕ characteristics curve (Typical)
(Starting from cold)

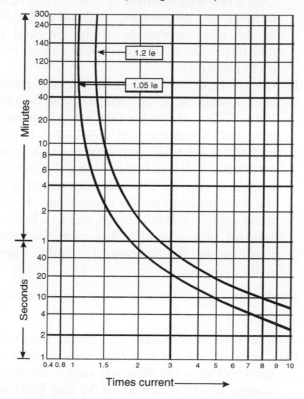

Figure 7.17
Operating characteristics of a bimetal relay

Motor startup or running problems are listed below:

1. The motor starts running as the start pushbutton is pressed, but stops as soon as the start pushbutton is released.
2. The motor starts running and trips 2 min after the start pushbutton is released.

Let us assume the main circuit fuses are not blown.
The following is the solution to the above-listed problems:

- Since the motor starts running as the start pushbutton is pressed, it indicates that the main contactor (K1) coil gets control supply when the circuit is completed on pressing the start pushbutton. However, the motor trips as soon as the start pushbutton is released.

- In the control circuit, as soon as the main contactor is switched on, the NO contact parallel to the start pushbutton contact must also close and hold the control circuit on until the stop pushbutton is pressed or the overload relay trips and its NC contact is open.

To troubleshoot the problem, perform the following steps:

(i) Check the control supply (L1), check control voltage between L1, and neutral (N).
(ii) Check the control fuse (F3) with the multimeter. If the control fuse (F3) is blown, change the fuse and restart the motor, the motor must start if control fuse (F3) blown was the only problem.

(iii) If the control fuse is OK, check whether the overload relay has tripped. Check this with the help of a multimeter. Check voltage between the neutral terminal and the outgoing terminal of the overload relay contact, connected to the stop pushbutton. If the overload relay has not tripped and the multimeter shows that the control voltage between the two points is OK, then go to point (iv).

(iv) Check the control voltage at the stop pushbutton outgoing terminal to the start pushbutton. If the voltage is OK, then go to point (v).

(v) If two NO contacts are connected in parallel to each other, and the motor runs only when the start pushbutton is pressed, it indicates that, the NO contact of the main contactor must close as soon as the main contactor is switched on. It also indicates that the contactor that holds the control circuit on is not closing. The wires connected in parallel from the NO contact to the start pushbutton NO contact may be closed, or the NO contact of the main contactor is not closed, due to a faulty contact. To confirm this, take a loop of insulated wire, and short the contact K1; if the motor starts, it confirms that the NO contact is faulty. Change the NO contact block of the main contactor.

If the motor runs and trips after 2 min, to troubleshoot this perform the following steps:

(i) Check the control supply (L1), check control voltage between L1 and neutral (N).

(ii) Check the control fuse (F3) with the multimeter. If the control fuse (F3) is blown change the fuse and restart the motor. The motor must start if the control fuse (F3) blown was the only problem.

(iii) If the control fuse is OK, check whether the overload relay has tripped. Check this with the help of a multimeter, by checking the voltage between the neutral terminal and the outgoing terminal of the overload relay contact connected to the stop pushbutton. If the overload relay has tripped, you will not get control voltage between the two points. Reset the overload relay and look into the reasons for which the motor tripped on overload.

If there is a voltage between the two points, look for a loose contact or loose wiring at the subsequent contacts in the control circuits.

7.13   Troubleshooting strategies

Strategies for troubleshooting of control circuits and 'Ladder Logic Circuits':

1. It is important to have the control circuit drawings, details of devices, their interconnection and interlocking while troubleshooting the control circuits. To troubleshoot a machine or equipment problem, it is good to have the 'Manufacturer's Operation and Maintenance Manual', as well as the 'Troubleshooting Instructions'.

2. 'Block Interlocking Diagram' and 'Control Sequences' of the equipment/machine operations should be available during troubleshooting.

3. Drawings and details of the power circuit of the equipment or the machine, control devices, contactors, timers, counters, safety, and protection devices, etc. are needed for troubleshooting the root cause.

4. Appropriate test and measurement instruments required for testing the control and power circuit of the equipment, or the machine must be available.

5. Switch OFF the main power supply to the equipment/machine and switch control supply ON, to avoid any mishaps or accidents while troubleshooting control circuits because of the sudden starting of the equipment.

6. As control circuits are different from equipment-to-equipment and machine-to-machine, it is not be possible to formulate a single or common strategy for troubleshooting control circuits. However, exemplary/standard engineering and trade practices must be followed while troubleshooting the control circuits.

7.13.1   General document checklist for troubleshooting

✓ Control circuit drawings
✓ Manufacturers operations and maintenance manuals and troubleshooting instructions
✓ Block interlocking diagram and control sequences involving the equipment/ machine
✓ Drawings and details of power circuits of the equipment/machine
✓ Details of devices, control devices, contactors, timers, counters, safety/protection
✓ Power circuit of the equipment or the machine.

Consider the example of a drill machine and conveyor table in Figure 7.18.

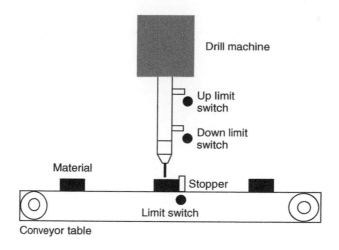

Figure 7.18
Sequence start stop

The sequence is as listed below:

1. The conveyor table should run as long as material strikes the conveyor limit switch that is provided.
2. The 'Stopper and Drill' should go up and down until it strikes the 'Down' limit switch. This makes a hole in the stationary material for 1 s. Again, the 'Drill' should go up until it strikes the 'Up' limit switch.

3. Then the conveyor starts again until the next material strikes the conveyor limit switch.
4. Outputs from PLC can be configured for conveyor start/stop, drill machine start/stop, drill up/down and stopper up/down.
5. Inputs to the PLC are the conveyor limit switch and drill machine up/down limit switch, conveyor table on/off.

7.14   Ladder logic design exercise

Prepare a 'PLC Ladder Logic Control Circuit' from the above example.

Appendix A

Units and abbreviations

Every science has its own unique language, and Electrical and Electronics Engineering is no exception. In this manual, some of the most common definitions and abbreviations are given in the text, where they appear first. To learn more definitions and abbreviations, please consider buying an electronic dictionary. They are available in most specialized bookshops and in most cases, are an excellent value for money.

Most countries in the world, including Australia, use the international system of units, called Systeme International d'Unites, and known simply as SI. The common electric terms, according to SI, have the symbols and units in Table A.1.

Item	Symbol	Unit	Symbol
Current	I	Ampere	A
Voltage	V	Volts	V
Charge	Q	Coulomb	C
Energy	W	Joule	J
		Watt Hour	Wh
Power	P	Watt	W
Resistance	R	Ohm	Ω
Resistivity	ρ	Ohm.Meter	Ωm
Reactance	X	Ohm	Ω
Inductance	L	Henry	H
Capacitance	C	Farad	F
Frequency	f	Hertz	Hz
Time	t	Second	s

Table A.1
Common electronic units

Appendix B

Troubleshooting

B.1 Component testing

A completed study of the manual would lead to mastery in the following areas:

- Testing diodes, SCRs, and TRIACs
- Testing BJTs, JFETs, and MOSFETs
- Using Ohms and Kirchhoff's laws to troubleshoot biased BJTs and FETs.

B.1.1 Testing diodes

As was discussed in Chapter 2, the diode is a semiconductor device, which conducts direct current in one direction only. In other words, the diode exhibits a very low resistance when it is forward-biased and an extremely high resistance when it is reverse-biased. In Chapter 7, it was mentioned that an ohmmeter applies a known voltage from an internal source (batteries) to the measured resistor. Theoretically, this voltage can reach 1.5 or 3 V. The diode requires a voltage of 0.7 V to become forward-biased. Therefore, if the positive test lead of the ohmmeter is connected to the anode and the negative test lead of the ohmmeter is connected to the cathode, the diode becomes forward-biased. In this case, the ohmmeter reads a very low resistance. If the test leads are reversed with respect to the anode and the cathode, the diode becomes reverse-biased. Then, the ohmmeter reads a very high resistance. Therefore, an ordinary ohmmeter can be used to test a diode.

Most DMMs have a diode-test function. It is marked on the select switch with a small diode symbol. When the DMM is set to the diode-test mode, it provides a sufficient internal voltage to test the diode in both directions. Figure B.1 illustrates the testing procedure of a diode. The positive test lead of the DMM (in red color) is connected to the anode, and the negative test lead of the DMM (in black color) is connected to the cathode. If the diode is in a good working order, the multimeter should display a value in the range between 0.5 and 0.9 V (typically 0.7 V). Then the test leads of the DMM are reversed with respect to the anode and the cathode. As the diode in this case appears as an open circuit to the multimeter, practically all of the internal DMM voltage will appear across the diode. The value on the display depends on the meter's internal voltage source and it is typically in the range between 2.5 and 3.5 V.

A defective diode appears either as an open circuit or as a closed circuit in both directions. The first case is more common and it is mainly caused by an internal damage of the PN junction due to overheating. Such a diode exhibits a very high resistance when it is both forward-biased and reverse-biased. On the other hand, the multimeter reads 0 V in both directions, if the diode is shorted. Sometimes a failed diode may not exhibit a complete short circuit (0 V) but may appear as a resistive diode, in which case the meter reads the same resistance in both directions (for example, 1.5 V). This is illustrated in Figure B.2.

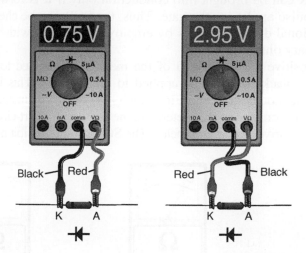

Figure B.1
Properly functioning diode

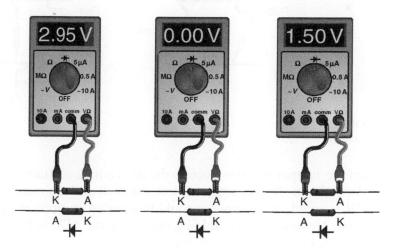

Figure B.2
Defective diodes

As was mentioned earlier, if a special diode-test function is not provided in a particular multimeter, the diode still can be checked, by measuring its resistance in both directions. The selector switch is set to Ohms. When the diode is forward-biased, the meter reads from a few hundred to a few thousand ohms. The actual resistance of the diode normally does not exceed 100 Ω, but the internal voltage of many meters is relatively low in the Ohms range and it is not sufficient to forward-bias the PN junction of the diode completely. For this reason, the displayed value is higher. When the diode is reverse-biased, the meter usually displays some type of out-of-range indication, such as 'OL', because the resistance of the diode in this case is too high and cannot be measured from the meter.

The actual values of the measured resistances are unimportant. What is important, though, is to make sure that there is a major difference in the readings, when the diode is forward-biased and when it is reverse-biased. In fact, this is all that is important to note, for this indicates that the diode is working properly.

B.1.2   Testing SCRs

As was discussed in Section 2.4.1, the SCR is a diode with an additional gate terminal. The SCR can be brought into conduction only if it is forward-biased and if it is triggered from a pulse applied to the gate. Thus, the SCR can be checked in a similar manner to the conventional diode, which is by employing a DMM with a diode-check function, or with an ordinary ohmmeter.

The positive (red) test lead of the meter is connected to the anode of the SCR and the negative (black) test lead is applied to the cathode. This is illustrated in Figure B.3. The instrument should show an infinite high resistance. A jumper can be used to trigger the SCR. Without disconnecting the meter, use the jumper to short-circuit the gate terminal of the SCR with the positive lead of the meter. The SCR should exhibit a great decrease of resistance.

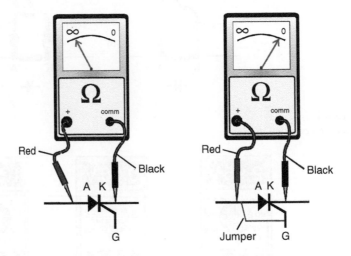

Figure B.3
Testing the SCR

When the jumper is disconnected, the device may continue to conduct or may turn off. This depends on the properties of both the SCR and the meter. If the holding current of the SCR is small, the ohmmeter could be capable of supplying enough current to keep it turned on. However, if the holding current of the SCR is high, the device will turn off upon the disconnection of the jumper.

Some high-power SCRs may have an internal resistor connected between the cathode and the gate. This resistor prevents the SCR from triggering due to the small interference surges. A maintenance technician, who is not aware of the existence of this resistor, may mistakenly diagnose such an SCR as being leaky between the cathode and the gate. The resistor's value can be measured with an ohmmeter during the test.

B.1.3   Testing TRIACs

The TRIAC actually consists of two SCRs connected in parallel and in opposite directions; therefore, the procedure for testing a TRIAC is essentially the same as the testing of an SCR. The positive test lead of the meter is connected to MT2 and the negative test lead is applied to MT1. When the gate is open, the ohmmeter should indicate an infinite resistance. Then, similar to the SCR testing procedure, a jumper is used to touch the gate terminal to MT2 (a positive triggering pulse is applied to the gate). The TRIAC should exhibit a great decrease in resistance. This indicates that one of the SCRs in the pair functions properly.

Then the test leads of the ohmmeter are reversed with respect to the anode and the cathode. Again, if the gate is open, the ohmmeter should exhibit an out-of-range resistance. Using the jumper, the gate terminal is briefly touched to MT2 (a negative triggering pulse is applied to the gate). The resistance of the TRIAC greatly decreases, which indicates the proper functioning of the second SCR in the pair. This procedure is illustrated in Figure B.4.

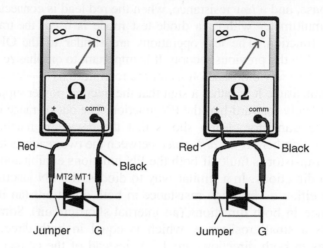

Figure B.4
Testing the TRIAC

B.1.4   **Testing BJTs**

As discussed in Chapter 2, the BJTs are devices, consisting of three layers of semi-conductive material and can be either of pnp or npn type. Therefore, each transistor can be represented as a combination of two diodes, connected as shown in Figure B.5. The equivalent base of pnp-type transistors appears as connected to the cathodes of both the diodes. If the transistors are of the npn type, the equivalent base appears as connected to the anodes of both the diodes. The two remaining terminals of the diodes represent the emitter and the collector. Both the PN junctions of the transistor are tested separately as two independent diodes. If both of them show no defects, the transistor is working properly.

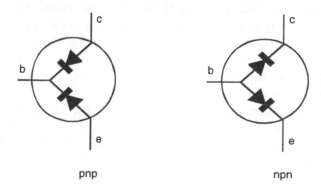

Figure B.5
A transistor, represented as two diodes

The diode-test function of a DMM can be also used to test the transistors. Let us assume that a pnp-type transistor has to be tested. The negative test lead (black) of the multimeter is applied to the base of the transistor. The positive test lead (red) is applied first to the emitter and then to the collector. In this arrangement, both the junctions will be forward-biased when tested. The DMM should read a low resistance in both cases. Then the red test lead is applied to the base of the transistor instead of the black one. The procedure is repeated. Both the PN junctions are now reverse-biased, when tested. The multimeter reads high resistance in both cases. The procedure for testing the npn transistors is identical. The difference is that the DMM will now read a *high* resistance, when the black lead is applied to the base, and a *low* resistance, when the red lead is connected to it.

If a multimeter without a diode-test mode is used, the transistor can be tested with the OHMs function. The test operations are similar to the OHMs function diode checking, described in the previous section. It is important to emphasize again, that the reading of a few hundred to a few thousand ohms for the forward-bias condition does not necessarily indicate a faulty transistor. It is rather a sign that the internal power supply of the meter is not sufficient to completely forward-bias the PN junction. The out-of-range indication for reverse-biasing of the same transistor clearly shows that the device is functioning properly. The important consideration here is the *difference* between the two readings and not their actual value.

The transistor is faulty if both the PN junctions exhibit approximately the same resistance in both directions. In a similar way to diodes, the PN junctions of the defective transistors exhibit either a very high resistance in both directions (an internal open-circuit), or a zero resistance in both directions (an internal short-circuit). Sometimes the faulty PN junction exhibits a small resistance, which is equal in both directions. For example, the meter readings in both directions are 1.2 V instead of the correct 0.7 V and the 2.9 V readings, respectively. In this case, the transistor is defective and has to be discarded.

Most DMMs are capable of measuring the current gain of the transistor bDC. The three transistor terminals are placed in special slots, marked E, B, and C, respectively. Then, a known value of IB is applied to the transistor and the respective I_C is measured. As you know, the ratio I_C/I_B is equal to bDC. Though this is a convenient and quick method to check the transistor, one should be aware that some DMMs measure the value of bDC with a low accuracy. The specifications of the DMM have to be checked before relying on the measured value of the current gain. Some testers have the useful feature of an in-circuit bDC check. Here there is no need to disconnect the suspected transistor from the rest of the circuit and it can be tested directly on the PCB.

B.1.5   Troubleshooting biased BJTs

Sometimes the transistor itself may not be faulty, but due to faults in the external circuitry, it may not operate correctly. For example, a cold junction on the transistor base terminal effectively isolates the base from the rest of the circuit. Therefore, the bias voltage on the transistor is 0 V, which will drive it into a cutoff. When checking such a transistor from the component side of the PCB, it will appear to be functioning correctly. Yet, the signal is not present at the output.

To better understand how to troubleshoot a biased BJT, consider the amplifier stage example shown in Figure B.6. It is built on the transistor 2N3946. According to the data sheets, bDC for this transistor is in the range of 50–150. Therefore, we can assume that bDC for the specified transistor is 100. The bias voltages are chosen $V_{BB} = 3$ V and $V_{CC} = 9$ V. Performing some simple calculations, we can determine that:

$$V_{BE} = 0.7 \text{ V}$$

$$I_B = \frac{3 \text{ V} - 0.7 \text{ V}}{56 \text{ K}\Omega} = \frac{2.3 \text{ V}}{56 \text{ K}\Omega} = 41.4 \text{ }\mu\text{A}$$

$$I_C = \beta_{DC} I_B = 100(41.1 \text{ }\mu\text{A}) = 4.1 \text{ mA}$$

$$V_C = 9 \text{ V} - I_C R_C = 9 \text{ V} - (4.1 \text{ mA})(1 \text{ K}\Omega) = 4.9 V$$

The voltages and the component values are specified in the Figure B.6. All the measured voltages are with respect to the ground. If the circuit operates correctly, the following voltages should be measured: +0.7 V in point A, +4.9 V in point B, and 0 V in point C.

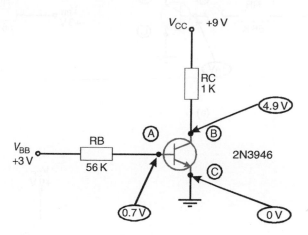

Figure B.6
Troubleshooting a single amplifier stage

First, the transistor has to be checked. If the transistor is not defective, the PCB has to be inspected visually for mechanical defects, burned components, and badly soldered joints. Finally, the voltages on the transistor terminals have to be measured.

Three typical abnormal conditions may occur due to faults in the external circuitry. They are illustrated in Figure B.7. Measuring the voltages on the transistor terminals can help to more effectively detect these faults. If the voltage at point B is only several mV instead of the normal +0.7 mV, then this is an indication that the base of the transistor is open (Figure B.7(a)). The soldered joints at the base of the transistor and at RB have to be checked. The value of the RB has to be measured. Any external circuitry, leading to the base of the transistor has to be inspected for badly soldered joints and for components that are out of tolerance.

If the meter reads a few mV on the collector terminal (point B) it is an indication that the collector is not connected to the rest of the circuitry (Figure B.7(b)). At the same time, the voltage on the base terminal should be around 0.7 V, as the base-emitter PN junction is forward-biased. The soldered joints on the collector and the collector resistor to the PCB have to be inspected. The value of RC has to be measured. Any component, connected to the collector resistor, has to be checked.

If there is an open ground connection, the symptoms are as follows: +3 V at the base terminal and +9 V at the collector terminal, as there is no collector and no emitter currents (Figure B.7(c)). The voltage measured at the emitter is +2.5 V or more. This occurs because the internal resistance of the measuring voltmeter provides a forward current path. It flows from VBB, through RB, the base-emitter junction and through the measuring voltmeter to the ground. Thus, the voltmeter registers the voltage drop across the PN junction. The soldered joint on the emitter has to be checked. All external circuitry connected to the emitter also has to be checked and tested.

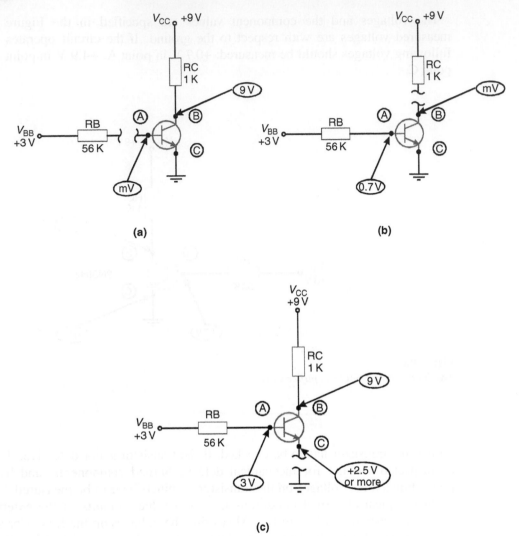

Figure B.7
Typical abnormal conditions in a biased BJT: (a) Open base; (b) Open collector; (c) Open emitter

B.1.6   Testing FETs

FETs are more difficult to test than BJTs. Before testing a FET, it must be ascertained if the transistor is a JFET- or a MOSFET-type. Thereafter, it has to be clarified whether it is a p-channel or an n-channel device. JFETs can be tested with an ordinary ohmmeter.

Figure B.8 depicts an equivalent circuit of a JFET. It appears to the ohmmeter as two diodes connected in series between the drain and the source. The polarity of the diodes is inverted. The gate terminal is taken from the midpoint between them. In the case of an n-channel type, the gate is connected to the anodes of both the diodes. If the transistor is a p-channel type, the gate is connected to the cathodes of both the diodes. The insulation layer of SiO_2 appears to the ohmmeter as a resistor connected between the drain and the source in parallel to both the diodes.

Therefore, the JFET transistors can be checked using an ohmmeter, by testing the PN junctions between the gate and the drain on one side and the gate and the source on the other. If the JFET is in good working order, both PN junctions should behave as ordinary diodes, exhibiting a high resistance in one direction and a low resistance in the other. Then the resistance between the drain and the source is measured. The meter should indicate some amount of resistance, which depends on the JFET properties.

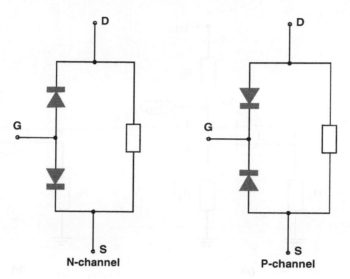

Figure B.8
A JFET, represented with two diodes and a resistor

In a best-case scenario, testing MOSFETs with an ohmmeter is a very difficult task. This is so because a very thin layer of metal oxide insulation separates the gate junction and the channel. This property of the MOSFET ensures extremely high input impedance of the device, but makes it vulnerable to permanent damage, even when minimal static voltages are built up at the transistor terminals. In fact, a MOSFET can be easily damaged even when it is lightly touched with a finger. For this reason, MOSFETs come in packages that provide an electrical connection between all terminals, which prevents the static voltages from building up.

MOSFETs can be tested very carefully with a low-voltage ohmmeter, set to the highest possible range. D-MOSFETs that are in a good working order, exhibit some continuity between the source and the drain. However, there should be no resistance, between the gate and drain and the gate and source terminals. E-MOSFETs that are working properly show no continuity between any of the terminals.

B.1.7   Troubleshooting-biased JFETs

It is not a recommended practice to unsolder a FET transistor in order to test it. After the visual inspection for damaged components or badly soldered joints, the voltages on the drain and the source have to be measured with respect to the ground.

A typical faulty symptom is the drain voltage, which is nearly equal to the power supply voltage (Figure B.9(a)). This condition occurs when the drain current is zero and therefore there is no voltage drop across R_D. The following faults may be the cause:

- Dry joint at R_S (R_S appears as an open circuit)
- R_S is faulty or it is out of tolerance
- Dry joint at R_D (R_D appears as an open circuit)
- R_D is faulty or it is out of tolerance

- Dry joint at the source terminal
- Dry joint at the drain terminal
- Internal JFET open-circuit between the source and the gate terminals.

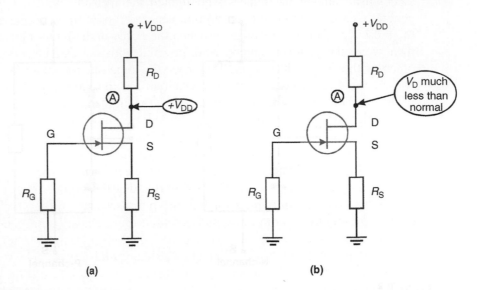

Figure B.9
Typical abnormal conditions in a biased JFET

Another typical faulty symptom is a drain voltage that is much less than the normal value (Figure B.9(b)). This condition occurs when the drain current is at a far higher level than normal, in which case there is a high voltage drop across R_D. The following faults may cause this to happen:

- Dry joint at R_G (R_G appears as an open circuit)
- R_G is faulty or it is out of tolerance
- Dry joint at the gate terminal
- Internal JFET open circuit at the gate terminal.

Some faults are very difficult to troubleshoot. One such example is an internally opened gate in a zero-biased D-MOSFET. After the fault occurs, the gate to the source voltage remains the same (0 V). For this reason, the drain current does not change its value and the bias appears to be normal. In general, troubleshooting FETs is a much more difficult task and requires more skills and experience than troubleshooting BJTs.

B.1.8   Troubleshooting op-amps

Op-amps are complex and sophisticated devices and are subject to several internal failures while in operation. However, the operational amplifier as such cannot be tested. Should there be an internal problem, it is not possible to troubleshoot and fix it. Therefore, if the op-amp fails, the only option is to replace it.

Usually there are only a few external components in the op-amp circuits. A typical circuit consists of an input resistor, a feedback resistor, and a potentiometer for an offset voltage compensation. If the circuit malfunctions, the external components have to be checked first. There could be dry joints, or the components may be burnt, or out of tolerance. If this is not so then, the contacts on the op-amp itself have to be checked. It is possible that some of them are faulty. Finally, if everything else appears to be in good working order, but the circuit is still non-operational, it has to be assumed that the amplifier itself is faulty. In this case, the op-amp is simply replaced as one would replace a resistor, a transistor, or any other component.

Some typical faults in op-amp circuits are given below:

- *Power supply voltage*: This is the first thing that should be checked (as is the case with troubleshooting any other circuit). A proper supply voltage and a ground must be present. It should be remembered that the level of the power supply is quite critical for most ICs.
- *Open feedback resistor*: This fault results in a severely clipped output voltage, as the op-amp operates at its maximum voltage gain (i.e., the circuit appears as an open-loop amplifier).
- *Shorted feedback resistor*: In this case, the output signal has the same amplitude as the input signal.
- *Open input resistor*: In the case of an inverting amplifier, there is no signal at the output, as there is no signal at the input. In the case of a non-inverting amplifier, the gain is equal to 1, and the output voltage follows exactly the input voltage. In other words, the amplifier will act as a voltage-follower.
- *Incorrectly adjusted potentiometer*: This fault results in clipping only the positive, or only the negative peak of the output voltage.

B.1.9   Summary

Most DMMs provide special functions for testing diodes and BJTs. However, if such functions are unavailable, most electronic components can be tested with an ordinary ohmmeter. If a diode is in a good working order, the ohmmeter readings should change from high to low (and the vice versa) every time the test leads are reversed with respect to the anode and the cathode. The SCR is tested in a similar way. The difference is that in addition a jumper is used between the gate and the anode to trigger the SCR. When the SCR is triggered, its resistance drops significantly from high to low. TRIACs are tested as SCRs, but in both directions (i.e., the test leads of the meter are reversed and the procedure is repeated). BJTs are treated as two diodes, connected in series. Each equivalent diode is tested independently. FETs are more difficult to test. Special care must be taken not to damage the device due to the static charge build up. JFET transistors can be represented as two diodes connected in series, with an additional resistor connected in parallel across them. Both equivalent diodes are tested independently. The value of the resistor is also measured. To find out faults in biased transistor circuits, initially, the approximate voltages on each transistor terminal are calculated. Then the voltages are measured. Any deviations from the calculated values are analyzed logically, which essentially leads to finding and fixing the problem. Op-amps cannot be tested, as the other devices. All external components and the soldered joints have to be checked and if the circuit still does not operate properly, the op-amp has to be replaced.

B.1.10   Quiz

1. A diode, tested with a DMM, exhibits 1.1 V in one direction and an out-of-range indication ('OL') in the other. Is the diode faulty?
2. A diode, tested with a DMM, exhibits the same value of 1.1 V in both directions. Is the diode faulty?
3. Does it make any difference if the jumper is connected to MT1 or MT2 during the test of a TRIAC?

4. Can you describe how to identify the type of an unknown transistor (pnp or npn) and its terminals (E, B, C), using just an ordinary ohmmeter?
5. The power supply of a one-stage BJT amplifier is 12 V. What voltage would you expect to measure at the collector?
6. A single-stage, small-signal amplifier is built using a JFET. The voltage measured at the drain is 0 V. What could be the cause of the fault?
7. A single-stage, small-signal amplifier is built using a BJT. Assume that the input signal is a sinusoid with a peak value of 0.5 V and the power supply is 12 V. There is a dry joint on the emitter. What is the form of the output voltage?

B.2 Troubleshooting techniques

A completed study of this chapter leads to a mastery in the following:

- Determine the relationship between the symptom and the cause of a problem
- Troubleshoot common circuits, using accepted techniques
- Conduct a fault analysis.

B.2.1 Symptom and the cause

The troubleshooting of electronic circuits involves three steps, which should be done in a specific order. The first step is to identify the defect in the circuit. The second step includes fault analysis and determination of the possible causes. The third step is fixing the problem.

First, it is important to identify the problem. To do so, symptoms have to be recognized in the defective circuit. A defective circuit can be defined as one, where the output parameters are incorrect, although the input parameters are correct. For example, the input signal of the amplifier, depicted in Figure B.10(a) is correct, but there is no signal at the output. In this case, the symptom is lack of voltage at the output.

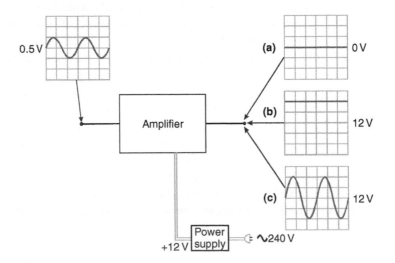

Figure B.10
Identifying the symptoms in a defective circuit

This particular symptom does not provide much information about the possible causes of the defect. The failure of various components in the circuit will result in the same symptom (zero voltage at the output). In other cases, a particular symptom points directly to a certain area where the fault is most likely to have occurred. For example, a DC voltage at the output, with the level equal to the supply voltage, indicates that there is a transistor in a cutoff condition, in the circuit (Figure B.10(b)). Starting from the stage that is closer to the output and going backwards, all transistors have to be checked for an internally open PN junction. The soldered joints and the values of the emitter resistors also have to be checked.

If the amplifier is not defective, the amplified signal appears at the output. The amplitude of the output signal is approximately equal to the value of the rectified power supply. The waveform has to be an exact amplified replica of the input signal, without any kind of distortion. This is illustrated in Figure B.10(c).

B.2.2   Troubleshooting techniques

Once the symptom is identified, the reasons that cause it have to be determined. The choice of which method to use depends on the circuit complexity, on symptoms, and on the personal preferences of the technician. The most common troubleshooting techniques are listed below:

- *Power check*:   It is amazing how many times a simple issue such as a blown fuse or a flat battery is the cause of a circuit malfunction. Initially, therefore, ensure that the power cord is plugged in and that the fuses are not blown. If the circuit is battery powered, make sure that the voltage level is acceptable. If a power supply rectifier is present, check the level of the voltage at the output and make sure that the circuit is powered with the correct polarity.

- *Visual inspection*:   This inspection is part of the so-called sensory checks. Sensory checks rely on the human senses to detect a possible fault. The visual inspection of the PCB is the simplest troubleshooting technique (which is very effective in many of the cases). The soldered joints have to be inspected thoroughly. If any doubts exist about the quality of a certain joint, it has to be re-soldered. The PCB has to be inspected visually for any burnt components. Sometimes, components that overheat leave a brownish mark on the board. They can be used as 'starting points' in the troubleshooting process and the reasons why they overheat have to be determined. It is bad practice simply to replace such components, without trying to find out what actually caused the component to overheat. In many cases, the reason is a faulty (or out of range) component near the failed component. It also has to be replaced.

- *Using a sense of touch*:   This is another sensory check. Overheated components can be detected by simply touching them. However, this check has to be performed with extreme caution. The circuit has to be turned off, and some time allowed for the large capacitors to discharge. Always touch the components with the right hand only. This is important because in the case of electric shock it is less likely that the current will pass through the heart. If possible, wear insulated shoes. In addition, care should be taken not to burn the fingers. Using the sense of touch is a very useful troubleshooting technique in circuits, where everything seems to work properly for a while, and then the circuit fails, due to overheating of a certain component. Identifying such components helps to detect the possible cause of the fault. Special freezing sprays are available, which allow instant freezing of components. If the circuit begins to operate properly immediately after the heated component is sprayed, this is an indication that this component is causing the circuit failure. Before replacing the component, further investigation is needed to determine what caused the overheating in the first place.

- *Smell check*:   When certain components fail due to overheating it is possible in most cases to detect a smell of smoke. This is usually the case, if the technician happens to be there at the time the accident occurred. If not, it is usually possible to detect the failed component by visual inspection afterwards.

- *Component replacement*:   This troubleshooting method relies mostly on the operator's skills and experience. Certain symptoms are an obvious indication of a particular component failure. This statement is especially true for an experienced electronic technician. For example, some TV service technicians can unmistakably identify the failed component in a TV set (even before opening it), by just briefly examining the symptoms. Component replacement is a good troubleshooting technique for an experienced electronics technician, as it saves a lot of time and money. Moreover, this technique guarantees the success of the repair, because if enough components are replaced, eventually the faulty one will be replaced too. However, it is recommended that the amateur technician initially applies some logical thinking to the troubleshooting process.

- *Signal tracing*:   This troubleshooting technique is not the most common one, but it is the most desirable, as it requires intelligent and logical thinking from the troubleshooter. This method is based on the measuring of the signal at various test points along the circuit. A test point in the circuit is the point, where the value of the voltage is known to the operator. This troubleshooting technique relies on finding a point, where the signal becomes incorrect. Thus, the operator knows that the problem exists in that portion of the circuit, between the point where the signal becomes incorrect, and the point where the signal appeared correct for the last time. In other words, the operator constantly narrows the searched portion of the circuit, until he finds what caused the fault. There are two basic approaches in conducting the signal tracing. In the first approach, the signal check starts from the input, checking consecutively the test points towards the output. The checks are carried out, until a point when an incorrect signal is found. The second approach is to start from the output and to work backwards towards the input in the same manner until a correct signal appears.

B.2.3   Fault analysis

Fault analysis requires a good theoretical knowledge and analytical thinking. It is not something which can be studied from books, but has to be acquired through constant troubleshooting and experimenting. The basic question in fault analysis is: 'What would the symptoms in the circuit be, if the component X is faulty?' For each specific application, there are no ready answers to this question. If there were, many books devoted to industrial electronics would be meaningless. However, there are certain rules, which can be adhered to, during the troubleshooting process. One of the tasks of this manual is to teach you some of these basic rules.

As an example, let us examine a bridge rectifier, to illustrate the process of fault analysis. The block circuit of a bridge rectifier that is working properly is shown in Figure B.11. It consists of a transformer, a rectifier, and a filter. The voltages, taken with an oscilloscope at each test point are depicted in the figure.

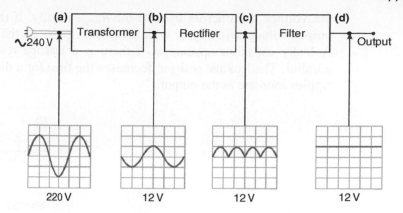

Figure B.11
A block diagram of a rectifier in good working order

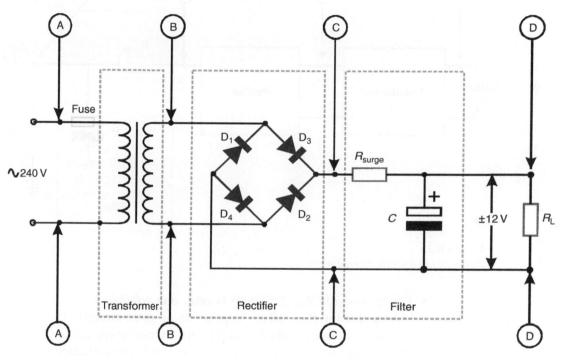

Figure B.12
A circuit diagram of the bridge rectifier

The circuit diagram of the same bridge rectifier is depicted in Figure B.12. A signal trace is conducted commencing from the output and working towards the input. An analysis of all possible faults in this circuit are given below:

- *Faulty capacitors* C: There are three possible problems. The capacitor could be shorted, opened, or leaky. If the capacitor is shorted, it effectively brings both terminals of the load resistor together and therefore the output voltage is zero. This is illustrated in Figure B.13(a). If the capacitor is open (Figure B.13(b)), it does not filter the output voltage supplied from the rectifier. The waveform of the voltage, at the output, remains the same as the waveform of the voltage, after the rectifier. Therefore, the waveforms at points C and D are identical. The only difference is that the amplitude of the voltage at the point D is smaller, due to the voltage drop across the resistor R_{surge}. Finally, if the capacitor is leaky the output voltage will appear with increased ripples on the output (Figure B.13(c)). A leaky capacitor appears as if there is a leakage resistor, connected to it in parallel. The leakage resistor decreases the time for a discharge, thus the voltage ripples increase at the output.

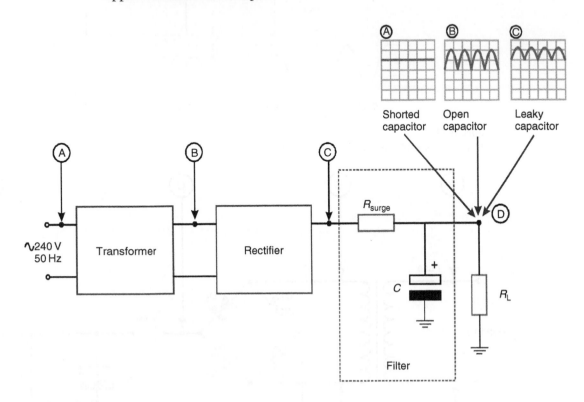

Figure B.13
Symptoms of a faulty capacitor

- *Faulty resistor* R_{surge}:   There is only one possible faulty condition, namely a blown resistor R_{surge} (R_{surge} appears as an open circuit). This occurs, when an excessive current flows through it. An excessive current flows through R_{surge} if the output terminals are short-circuited or if the capacitor is shorted. In both cases when R_{surge} blows, it brakes the circuit and prevents the diodes (which are more expensive than the resistor) from burning too. The output voltage in this case is zero. Before replacing R_{surge}, ensure that the capacitor, or the output terminals of the circuit, is not shorted and that the conductive paths of the PCB are not shorted out.
- *Shorted diode*:   shorted diode appears as a jumper between the points of the connection, as it conducts the current in both directions. Figure B.14 illustrates the current that flows in the circuit, when the diode D_4 is shorted out. During the positive half-period, the current flows through D_3 and D_4 as normal. The shortened diode exhibits a zero resistance in both directions and it appears for the circuit, as if it is simply forward-biased. Thus, the positive half-period appears as normal at the point C. However, during the negative half-period the picture changes. The current now flows through D_1 and D_4 instead of flowing through the rest of the circuit, because these two diodes, connected in series, provide a path of least resistance. Effectively the secondary winding is

short-circuited and an excessive current flows through it. Thus, the diode D_4 can be damaged quickly, due to overheating. The increase in the current in the secondary winding increases the current in the primary winding. If the circuit is properly fused, the fuse on the primary winding should blow. If this is not the case, the diode D_1 overheats (and even possibly burns) and the voltage at the test point C has the form shown in Figure B.14. Analytical thinking is required to analyze what happens in the circuit when some other diode shorts out, or when two or more diodes short out simultaneously.

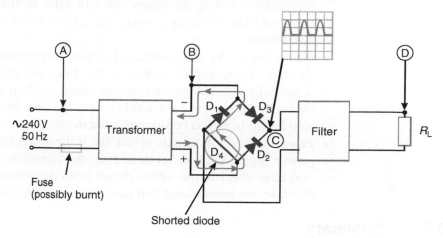

Figure B.14
Symptoms of a shortened diode

- *Open diode*: Let us assume that the same diode (D_4) is open. No current flows through an open diode in both directions. During the negative half-period, this diode appears to the circuit to be reverse-biased, and therefore it has no impact on the output voltage. However, during the positive half-period, the path for the current is broken and no voltage appears at the output. In other words, the circuit works as a half-wave rectifier. This can be detected by, the larger ripples in the output voltage. In addition, the frequency of the ripples is 50 Hz instead of 100 Hz. This is illustrated in Figure B.15. Similarly, the circuit can be analyzed for other open diodes.

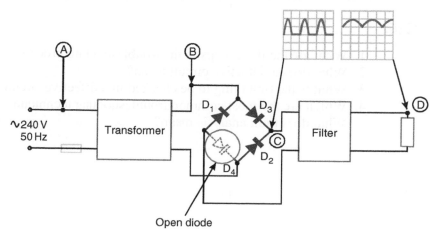

Figure B.15
Symptoms of an open diode

- *Faulty transformer*: This is not a common fault, though if the rest of the circuit appears in a good working order, the transformer has to be checked. Several faults are possible: the primary or the secondary windings can be open or partially shorted. If one of the windings is open, no voltage is applied to the rest of the circuit. This obviously results in 0 V at the output. If the primary winding is partially shorted, the turns ratio of the transformer is effectively increased. The voltage on the secondary winding is also increased; thus, the level of the voltage at the output of the circuit is higher. A partially shorted secondary winding decreases the turn ratio of the transformer. The voltage supplied to the rectifier is lower; thus, the level of the circuit output voltage is also lower.

- *Blown fuse*: As was mentioned earlier, this occurs when one of the diodes is shorted. Thus, before replacing the fuse, the diodes have to be checked. A partially shorted primary or secondary winding of the transformer can also increase the current to a level, where the fuse blows. Thus, the transformer also has to be tested before replacing the fuse.

- *Power supply*: It sounds trivial, but if the power cord is not plugged in, or the power supply knob is not turned on, the circuit obviously will not function. If you have checked the whole circuit from the output just to find out that this is the case, we recommend that next time you start from the input.

B.2.4   Summary

Troubleshooting of electronic circuits involves three steps, which should be performed in a specific order. The first step is to identify the defect in the circuit. A defective circuit can be defined as one, where the input parameters are correct, but which has incorrect output parameters. The second step includes a fault analysis and a determination of the possible causes. Here, various techniques can be applied. The most common are: the power supply check, sensory checks (visual, using the sense of touch, smell, etc.), component replacement, and signal tracing. Component replacement is a very effective troubleshooting technique and works well for experienced electronic technicians. Signal tracing involves measuring the voltage at test points, until a voltage with an incorrect value is found. This is the most desirable technique, as it requires a good knowledge of the theory, and some logical thinking. The third step of the troubleshooting process is fixing the problem. This can be done by component replacement, re-soldering a dry joint, etc.

B.2.5   Quiz

1. What are the three steps in the troubleshooting process?
2. What does a defective circuit mean?
3. What is the first thing to be checked in a defective circuit?
4. Which are the most common troubleshooting techniques?
5. What does fault analysis mean?

Appendix C

Low-voltage networks

C.1 Introduction

The low-voltage network is a very important component of a power system as it is at this level that maximum power is distributed and utilized by the end-user. Essential loads such as lighting, heating, ventilation, refrigeration, air-conditioning, etc. are generally fed at voltages such as 380 V, 400 V, 415 V, 500 V, 525 V, three-phase three-wire, and three-phase four-wire.

In the mining industry, heavy motor loads often require voltages as high as 1000 V. Due to the diverse nature of the loads, coupled with the large number of items requiring power, it is usual to find a bulk in-feed to an LV switchboard. This is followed by numerous outgoing circuits, of varying current ratings, in contrast to the limited number of circuits, at the medium-voltage level.

Large-frame (high current rated) air-circuit breakers (ACBs) are therefore referred to as 'incomers' from the supply transformer and MCCBs for all outgoing feeders. The downstream network generally consists of MCCBs of varying current ratings, and as the current levels drop, MCBs are used for compactness and cost saving.

C.2 Air-circuit breakers

These breakers are available in frame sizes ranging from 630 to 5000 A in 3- and 4-pole versions and are generally insulated for 1000 V. Rated breaking capacities of up to 100 kA rms symmetric to IEC947-2 are claimed at a rated voltage of 660 V.

Fixed and draw-out models are available and each unit invariably comes complete with a protection device, which in keeping with modern trends, is generally of the electronic type. This will be discussed in more detail later.

Typical total breaking times are of the order of 40–50 ms for short-circuit faults. Their operating speed is important, as ACBs are applied as the main incoming devices, to the low-voltage network, where they are subject to the highest fault levels determined by the supply transformer.

A typical construction of an ACB is shown in Figure C.1.

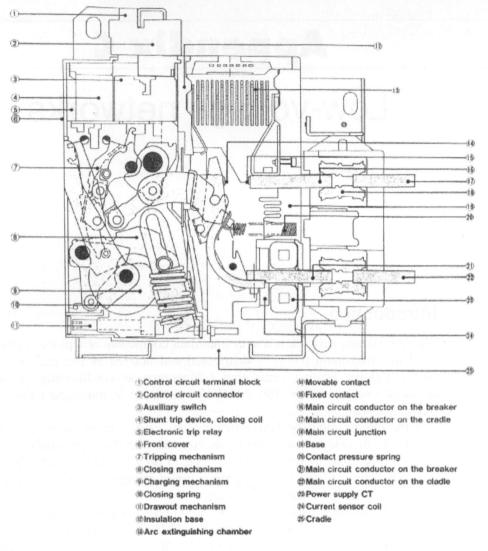

① Control circuit terminal block
② Control circuit connector
③ Auxiliary switch
④ Shunt trip device, closing coil
⑤ Electronic trip relay
⑥ Front cover
⑦ Tripping mechanism
⑧ Closing mechanism
⑨ Charging mechanism
⑩ Closing spring
⑪ Drawout mechanism
⑫ Insulation base
⑬ Arc extinguishing chamber
⑭ Movable contact
⑮ Fixed contact
⑯ Main circuit conductor on the breaker
⑰ Main circuit conductor on the cradle
⑱ Main circuit junction
⑲ Base
⑳ Contact pressure spring
㉑ Main circuit conductor on the breaker
㉒ Main circuit conductor on the cladle
㉓ Power supply CT
㉔ Current sensor coil
㉕ Cradle

Figure C.1
Typical internal construction of an-ACB

C.3   Molded-case circuit breakers

MCCBs are power switches with built-in protective functions used on circuits that require low current ratings. They include the following features:

- Normal load current – open and close switching functions
- Protection functions to automatically disconnect excessive overloads and to interrupt short circuit currents as quickly as possible
- Provide indication status of the MCCB either open, closed, or tripped.

Although many different types are manufactured, they all consist of five main parts:

1. Molded case (Frame)
2. Operating mechanism
3. Contacts and extinguishers
4. Tripping elements
5. Terminal connections.

C.3.1   Molded case

This is molded from resin/glass fiber materials, which combine ruggedness with high dielectric strength.

The enclosure provides a frame on which to mount the components, but more importantly, it provides insulation between the live components and the operator. Different physical sizes of case are required by the maximum-rated voltage/current and interrupting capacity, and these are assigned a 'frame size'.

C.3.2   Operating mechanism

This provides a means to open and close the MCCB via the handle. In passing from ON to OFF (or vice versa) the handle tension spring passes through alignment with the toggle link, and in doing so, a positive rapid contact-operating action is produced to give a 'quick break' or a 'quick make' action. This makes it independent of the human element, i.e., the force and speed of operating the handle.

The mechanism also has a 'trip free' feature, which means it cannot be prevented from tripping by holding the operating handle in the ON position. In other words, the protective contact-opening function cannot be defeated. Another important feature in addition to indicating when the breaker is ON (in the up position) or OFF (in the down position), the handle indicates when the breaker has tripped by moving midway between the extremes, as shown in Figure C.2.

To restore the service after the breaker has tripped, the handle must first be moved to the OFF position to reset the mechanism before being moved to the ON position.

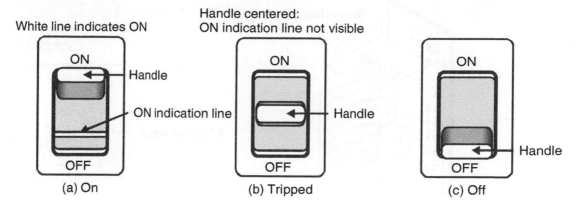

Figure C.2
Handle positions

C.3.3   Contacts and extinguishers

A pair of contacts comprises of a moving contact and a fixed contact. The instance of opening and closing impose the most severe duty. The contact materials must therefore be selected with consideration to the following three criteria:

1. Minimum contact resistance
2. Maximum resistance to wear
3. Maximum resistance to welding.

Silver or silver alloy contacts are low in resistance but wear rather easily. Tungsten or tungsten alloys are strong against wear due to arcing but rather high in contact resistance.

Thus, the contacts are designed to have a rolling action, containing mostly silver at the closing current-carrying points, and mostly tungsten at the opening (arcing) point (Figure C.3).

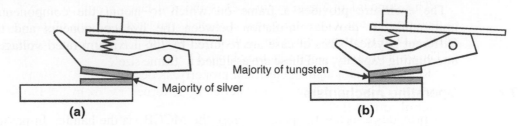

Figure C.3
Dual function contacts: (a) Closing; (b) Opening

In order to interrupt high short-circuit currents, large amounts of energy must be dissipated. This is achieved by using an arc shute (Figure C.4) that comprises of a set of specially shaped steel grids, isolated from each other, and supported by an insulated housing. When the contacts are opened and an arc is drawn, a magnetic field is induced in the grids, which draws the arc into the grids.

The arc is thus lengthened and chopped up into a series of smaller arcs which are cooled by the grids' heat conduction. Being longer, it requires far more voltage to sustain it and being cooler tends to lose ionization and extinguishes at the first current zero.

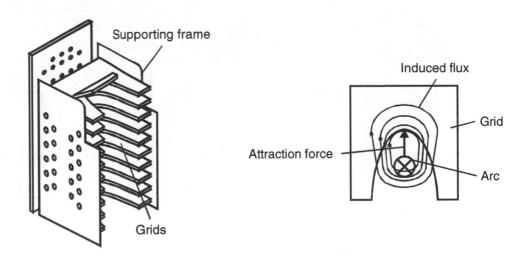

Figure C.4
The Arc shute

C.3.4 Tripping elements

The function of the trip elements is to detect the overload or the short-circuit condition and trip the operating mechanism.

Thermal overload

The thermal trip characteristic must be in close proximity to the thermal characteristics of cables, transformers, etc. To cover this overload condition, two types of tripping methods are available, namely Bimetallic and Hydraulic.

Bimetallic method

The thermal trip action is achieved by using a bimetallic element heated by the load current.

The bimetallic element consists of two strips of dissimilar metals bonded together. Heat due to an excessive current will cause the bimetallic element to bend because of the difference in the rate of expansion of the two metals. The bimetallic element must deflect far enough to physically operate the trip bar. These thermal elements are factory-adjusted and are not adjustable in the field. A specific thermal element must be provided for each current rating. A number of different variations on this theme are available as shown in Figure C.5.

The bimetal is temperature-sensitive and automatically re-rates itself with variations in the ambient temperature.

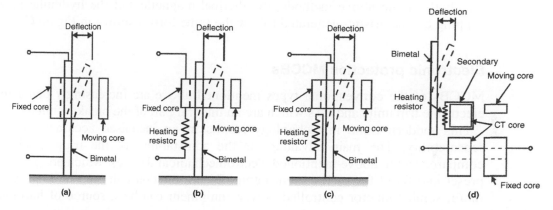

Figure C.5
Thermal tripping methods: *(a) Direct heating; (b) Indirect heating; (c) Direct–indirect heating; (d) CT heating*

Hydraulic method

For its operation, this device depends on the electromagnetic force, produced by the current flowing in a solenoid, wound around a sealed non-magnetic tube. The tube, filled with a retarding fluid, contains an iron core, which is free to move against a carefully tensioned spring.

For a normal load current, the magnetic force is in equilibrium with the pressure of the spring. When an overload occurs, the magnetic force exceeds that of the spring and the iron core begins to move, reducing the air gap in the tripping armature. Once the magnetic field is large enough, the armature closes to trip the mechanism. The time delay characteristic is controlled by the retarding action of the fluid. The concept is illustrated in Figure C.6.

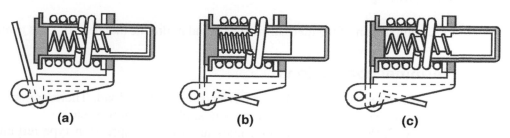

Figure C.6
Hydraulic tripping method: (a) Normal operation; (b) Long-delay tripping; (c) Instantaneous tripping

Short circuits

During short circuit conditions, response time of the thermal element is slow, therefore, a faster type of protection is required to reduce any damage. For this reason, a magnetic trip action is used in addition to the thermal element.

When a fault occurs, the short circuit current causes the electromagnet to attract an armature that unlatches the trip mechanism. This is a fast action and the only delay is the time it takes for the contacts to physically open and extinguish the arc. This is normally of the order of 20 ms – typically 1 cycle. In the hydraulic method, the current through the solenoid will be large enough to attract the armature instantaneously, irrespective of the position of the iron core. The interruption speeds for this type of breaker for short circuit currents are also less than 1 cycle (20 ms) and similar to the bimetallic type.

In both of the above methods, the thermal-magnetic and the hydraulic-magnetic, the tripping characteristics generated follow the same format shown in Figure C.7.

Electronic protection MCCBs

MCCBs of the conventional types mentioned above are increasingly being replaced by electronic trip units and CTs which are an integral part of the breaker frame.

This modern trend in technology results in an increased accuracy, reliability, and repeatability. The main advantage is the adjustability of the tripping characteristics, compared to the above-mentioned electromechanical devices, which are generally factory preset and fixed for each current rating. Discrimination can then be improved. Furthermore, semiconductor controlled power equipment can be a source of harmonics which may cause mal-operations.

Electronic protective devices detect the true RMS value of the current, thereby remaining unaffected by the harmonics.

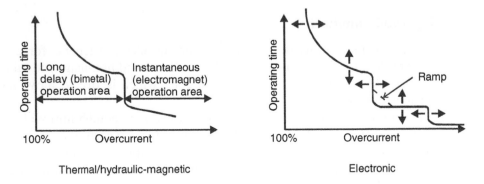

Figure C.7
Typical tripping characteristics

A comparison of the thermal-magnetic and hydraulic-magnetic types is given in Figure C.8.

C.3.5   Terminal connections

These connect the MCCB to a power source and a load. There are several methods of connection such as bus bars, straps, studs, plug-in adaptors, etc.

Up to 250–300 A whenever cables are used, compression type terminals are used to connect the conductor to the breaker. Above 300 A, stubs, bus bars, or straps are recommended to ensure reliable connections, particularly when using aluminum cables.

Item	Thermal-Magnetic Type	Hydraulic-Magnetic Type

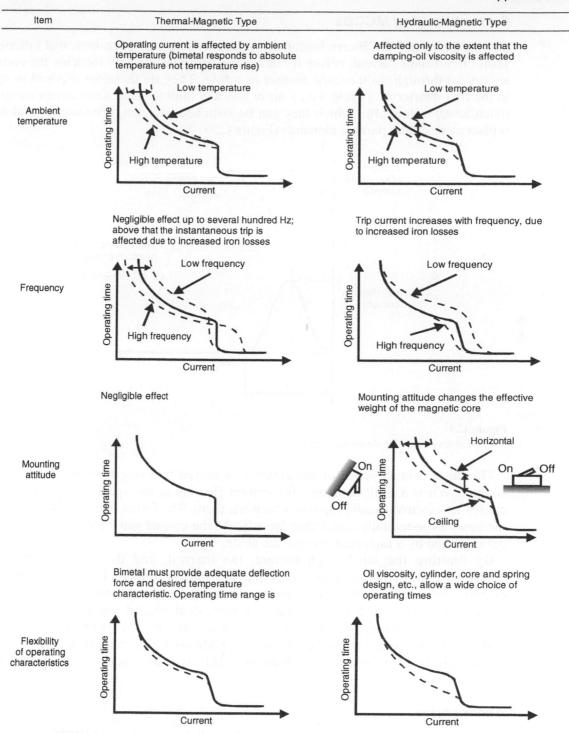

Figure C.8
Comparison of thermal-magnetic and hydraulic-magnetic types

Current-limiting MCCBs

Current-limiting MCCBs are essentially extremely fast-acting breakers, that interrupt the short circuit fault current, before it reaches the first peak, thereby reducing the current or energy let-through, in the same manner as a fuse. They are therefore required to operate in the first quarter of a cycle, i.e., 5 ms or less and limit the peak short circuit current to a much lower value, after which they can be switched on again, if necessary, without the replacement of any parts or elements (Figure C.9).

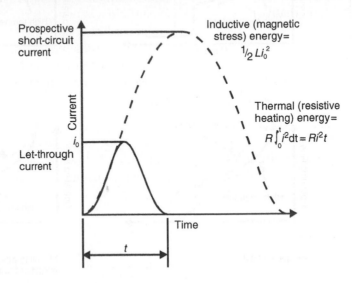

Figure C.9
Limited short circuit let-through current

This high contact speed of separation is achieved by using a reverse loop stationary contact. When a fault develops, the current flowing in the specially designed contacts causes an electrodynamic repulsion between them. The forces between the contact arms increases exponentially rather than linearly. As the contact gap widens, the arc is quickly extinguished by a high-performance arc shute.

By limiting the let-through current, the thermal, and the magnetic stresses on protected equipment such as cables and bus bars is reduced in the event of a short circuit. Provided combination series tests have been done and certified, this also permits the use of MCCBs with a lower short circuit capacity to be used at downstream locations from the current-limiting MCCB. This is known as 'cascading' and results in a more economical system. Additional care must be taken, to preserve the discrimination between the breakers. This will be discussed in more detail in Section C.4.

C.3.6   Accessories

The following accessories are available with all different makes of MCCB:

- Shunt trip coils
- Under-voltage release coils
- Auxiliary switches
- Mechanical interlocks
- Electrical closing mechanism (in higher ratings).

A typical MCCB is shown in Figure C.10.

When selecting an MCCB for an application it is important to ensure that the following ratings are correct:

1. Voltage rating
2. Current rating
3. Breaking capacity rating.

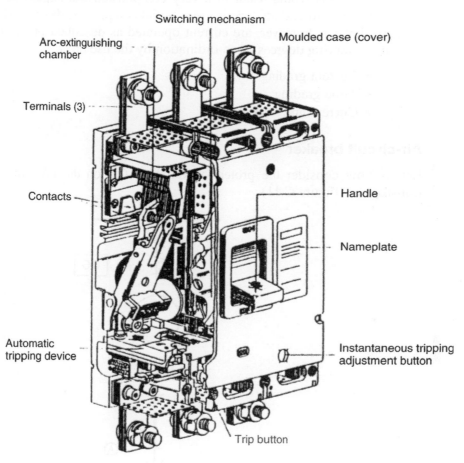

Figure C.10
Typical MCCB

C.4   Application and selective co-ordination

The basic theory of selective co-ordination is applicable for all values of the electrical fault current.

- *Milli-amperes*: Earth leakage protection
- *Hundreds of amps*: Overload protection
- *Thousands of amps*: Short circuit protection.

Earth leakage protection will be discussed later under Section C.5. In the short circuit situation, it is generally accepted that most short circuit currents that occur in practice fall below the calculated theoretical value for a three-phase bolted fault. This is so, because not all faults occur close to the MCCB (except when the supply cable is connected to the bottom of the MCCB). The resistance of the cable between the MCCB and the fault reduces the fault current. Most faults are not bolted faults – the arc resistance helps to reduce the fault current even further. For economic and practical reasons, it is not feasible to apply the same sophisticated relay technology, as used on the medium-voltage to low-voltage networks, as this would result in a very complicated and expensive system. The present system of using air and MCCBs is a successful compromise developed over many years.

These devices, however, are current operated as described previously, so it is possible to achieve varying degrees of co-ordination by the use of:

- Current grading
- Time grading
- Current and time grading.

Air-circuit breaker

Let us now consider the protection on the ACB on the LV side of the main in-feed transformer (Figure C.11).

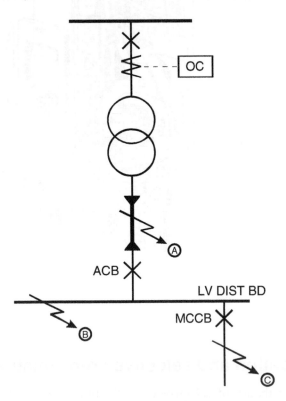

Figure C.11
LV ACB on transformer

Transformer overload condition

The thermal element on the ACB can be set, to protect the transformer against excessive overloading, as the same current that flows through the transformer flows through the ACB. Tripping this breaker removes the overload and allows the transformer to cool down. The transformer has not faulted – it is only being driven above its continuous design rating, which if allowed to persist for some time, will cook the insulation leading to an eventual failure. Checking the temperature indicators on the transformer allows the operator to have a clear indication of the problem. The transformer is still alive from the HV side so it has not faulted. It is purely on an overload condition.

It has often been a common practice to trip the transformer from the HV IDMT overcurrent relay, for an overload condition. With this approach, the operator does not know if the transformer has faulted or if it was just an overheating condition. When faced with such a decision, deciding to test the unit before switching it again could lead to an excessive downtime. In addition, the HV IDMT overcurrent relay (normal inverse) does not have the appropriate characteristic for an overload protection, as pointed out in Chapter 10.

Short-circuit protection

Short circuits at points A, B, and C must be considered. The fault currents will be the same, as there is virtually no impedance between them. The short circuit protection on the ACB should therefore be set with a short time delay, to allow the downstream MCCB to clear fault C. However, if the fault is on the busbar, the time delay should be short enough, to ensure a relatively fast clearance to minimize damage and downtime (Figure C.12).

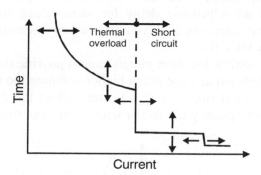

Figure C.12
ACB–adjustable protection tripping characteristics

Fault A will have to be cleared by the HV overcurrent relay in order to protect the cable, from the transformer to the LV switchboard. This in turn, should have a longer time delay, to co-ordinate with the LV ACB and also to provide, the discrimination for faults B and C. These requirements show the value, of specifying adjustable current pickups and time delays, for the protection devices on ACBs, most of which are available in an electronic form.

In addition, they also come equipped, with a very high set instantaneous overcurrent feature, having a fast fixed time setting of 20 ms, to cover 'closing-onto-fault' conditions.

C.4.1   Molded case circuit breakers

A reasonable degree of current grading can be achieved, between the two series-connected MCCBs by simply applying, a higher rated breaker, up-stream of a given unit. The extent of the co-ordination is shown on the following time-current characteristic curves (Figure C.13).

It will be noted that selectivity is obtained in the thermal overload and partial high current region, co-ordination being lost, above the short-circuit pickup current level, of the up-stream breaker.

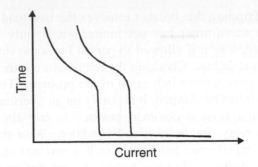

Figure C.13
Current co-ordination

For large consumers, the integrity of the supply is important. The ability of the up-stream breaker to hold in, under such fault conditions is enhanced, when it is equipped with an additional short time delay facility, provided by the modern electronic trip elements (Figure C.14).

Note: It is important to note that an MCCB does not have a short time fault current withstand rating unlike ACBs (which can usually withstand rated short circuit current for a duration of 1 s). A time difference of tripping of at least 0.3 s is necessary for proper co-ordination between an upstream and downstream breaker which is not provided in MCCBs as an adjustable delay for short-circuit trip. Because of this, accurate time co-ordination cannot be achieved between an upstream and downstream circuit when both of them use MCCBs.

Also, the setting for short circuit cannot provide the required current discrimination as the fault levels usually are much higher compared to the breaker current settings and also the point of fault does not significantly affect the fault current magnitude in industrial systems where usually LV feeder lengths are kept short.

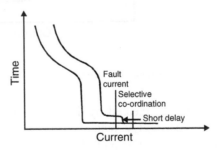

Figure C.14
Current-time co-ordination

C.4.2   MCCB un-latching times

Once triggered, MCCBs have an un-latching time which is dictated by the physical size and inertia of the mechanism. It stands to reason that the physically smaller, lower-rated breakers will have a shorter un-latching time than the higher-rated, larger up-stream breakers, thereby enhancing their clearing time.

Experience in practical installations of the fully rated breakers, has shown that unexpected degrees of discrimination have been achieved because of this. For current-limiting circuit breakers, where contact parting occurs, independent of the mechanism, the un-latching times do not have such an impact on their clearance times.

C.4.3 Fully rated systems

When time-delayed MCCBs are used to achieve an extended co-ordination, all the down-stream circuit breakers must be rated, to withstand and clear, the full prospective short circuit current, at the load side terminals. The installation of an MCCB and making cable connections to its terminals require caution. Please refer to figures C.15 and C.16.

Cautions for installation

Do not remove the rear cover

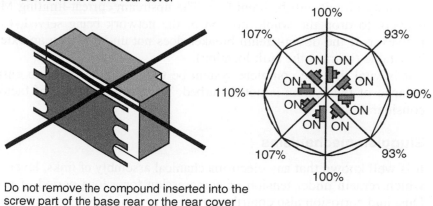

Do not remove the compound inserted into the screw part of the base rear or the rear cover

Change ratio of the rated current value according to the installation angle

Figure C.15
Cautions for installation

Cautions for connection

1. Take sufficient insulation distance

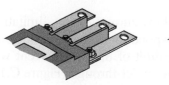

Take care, as the insulation distance may be insufficient according to the installation position of the connection conductor.
As some type are provided with insulation barriers, they should be used under reference top appendix table 6

2. Do not apply oil to the threaded parts

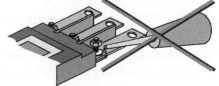

Do not apply lubrication oil to the threaded parts. Application of lubrication oil reduces the friction of the threaded part, so that loosening and overheat can be caused. In case of lubrication, even the standard tightening torque can produce excessive stress in the threaded part and thus breaking of the screw

1. Parallel conductors for all poles

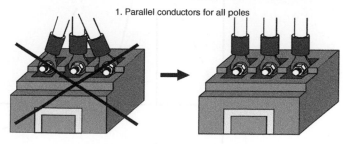

Install the connection conductors in parallel for all poles

Figure C.16
Cautions for connections

C.4.4   Cascading systems

This approach can be used, if saving on the initial capital cost is the overriding factor. This necessitates using a current-limiting breaker, to contain the let-through energy, thereby allowing lower-rated (hence less costly) breakers, to be used downstream. To achieve a successful co-ordination, careful engineering is required. This is so with regard to clearance and un-latching times, additionally to, the size and length of the interconnecting cables, together with an accurate calculation of the fault levels.

If the let-through energy is sufficient to cause the downstream breaker to un-latch, then the faulty circuit will be identified. The upstream current-limiting MCCB will also have tripped, to drop the whole portion of the network being served, by this main breaker. However, if the down-stream breaker does not un-latch then an extended outage time is inevitable (to trace the fault location).

It is vital that the complete system be tested and approved to ensure that the delicate balance of the system is not disturbed. There are a number of factors that need careful consideration.

Sluggish mechanisms

It is well known that any electromechanical assembly of links, levers, springs, pivots, etc. which remain under tension or compression for a long period of time, tend to 'bed in'. Dust and corrosion also contribute further, to retarding the operation, after long periods of inactivity. The combined effect could add a delay of 1–3 ms when eventually called into operation.

This additional delay has little effect on fully rated breakers which generally operate after one cycle (20 ms), but on the current-limiting MCCBs, which are required to operate in 5 ms, the additional 1–3 ms will have a significant impact on their performance. The increased energy let-through could have disastrous results, both for itself and the downstream breaker.

Point-on-wave switching

Most specifications and literature show current/energy-limitation, based on fault initiation occurring at a point-on-wave corresponding to the current zero. Should the fault occur at some other point on the wave, the di/dt of the fault current would be much greater than that shown, resulting in a higher energy let-through (Figure C.17).

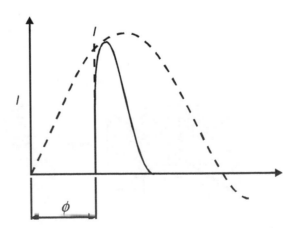

Figure C.17
Effect of point-on-wave fault occurrence

Service deterioration

'Qualification type tests' in most international specifications require the MCCB to successfully perform one breaking operation and one or two make-break operations. In practice, it is rare that the number of operations by a breaker under short-circuit conditions is monitored. This shortcoming is not critical on fully rated systems, as the protection of the downstream breakers is of no consequence.

However, in a series-connected cascading system, where the downstream breakers rely for their survival on the energy-limiting capabilities of the upstream current-limiting breaker, there is always the danger that replacement of the upstream device is overlooked. Therefore, there is a strong case for monitoring the number of operations.

Maintenance

For reasons stated above, any up-stream or downstream breakers in a cascade system must be replaced with identical breakers from the same manufacturer, in accordance with the original test approvals. This also applies to any system extensions. Any deviation could prove disastrous.

Incorrect replacement of the up-stream breaker could result in a higher energy let-through and longer operating times. Incorrect replacement of the downstream breakers may lead to a lower energy-handling capability, coupled with shorter operating times.

These conflicting requirements are such that, even experienced or well-trained technicians may be confused, unless they are fully conversant with the principle of the cascade system. There could be an even greater problem for the maintenance electrician and his artisan, in selecting a replacement device, which may often be dictated by availability.

Identification

In view of the problems of staff turnover and the possibility of decreasing skills, it becomes a stringent requirement that all switchboards carry a prominent identifying label, together with all relevant technical information, to ensure the satisfactory operation and maintainability of the cascaded or the series-connected systems.

General

Cascaded systems offer attractive savings in the initial capital expenditure. They, however, require a higher level of engineering for the initial design and extensions. Maintenance can be difficult, as the total knowledge and understanding of the system and all its components is required by all operating personnel. The consultant, contractor, or user is thus faced with the decision of choosing between two very different systems:

- A fully rated properly co-ordinated system
- A system based on cascaded ratings.

The first choice may have a slightly higher initial cost. The alternative offers some initial cost savings whilst sacrificing some system integrity, selectivity, and flexibility.

C.5   Earth leakage protection

In the industrial and mining environment, the possibility of operators making direct contact with live conductors is very remote. This is because the conductors are housed in specially designed enclosures, which are lockable and where only trained and qualified electricians are allowed access. Danger due to indirect contact (by touching the enclosure of an equipment having an earth fault) is addressed mainly by clearing the fault as quickly as possible. Since LV circuits are provided a metallic return path to the neutral point of source by proper earth bonds, it is ensured that the fault current magnitudes are high enough to be sensed and cleared by protection devices meant for isolating short-circuit faults such as fuses or circuit breakers.

The danger lies, however, when an earth fault occurs on a machine and because of poor earth-bonding, the frame of the machine becomes elevated to an unsafe touch potential, as illustrated in Figure C.18. This is an 'indirect' contact situation, which must be protected.

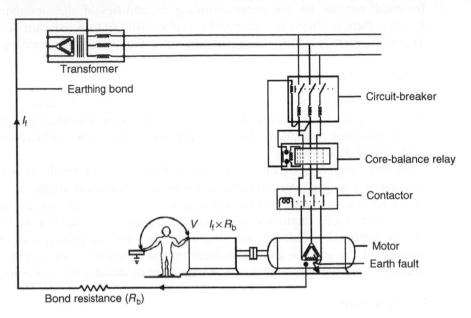

Figure C.18
Protection against indirect contact

Safety codes specify that in mining and industrial installations any voltage above the range of 25–40 V is considered unsafe. These figures are derived from the current level that causes ventricular fibrillation – 80 mA times the minimum resistance of the human body, which is in the range of 300 Ω (3 × safety factor) to 500 Ω (2 × safety factor). Please refer to Chapter 4, Section 4.2.

It would not be possible to utilize the sensitive domestic earth leakage devices (30 mA, 30 ms) in such applications because of the transient spill currents that occur during motor starting. Instantaneous tripping would occur and the machine would never get started. Tests have been carried out in coal mines to determine the maximum resistance that could occur on an open earth bond. The results were measured as 100 Ω. With 25 V specified as the safe voltage, then a current of 250 mA can be regarded as the minimum sensitivity level (derived from dividing 25 V by 100 Ω). This level was found to be stable for motor starting. It is, however, above the 80 mA fibrillation level of the heart, so speed is now of the essence, if we are to save a human life.

The earth leakage relays used in industrial applications should therefore operate in 30 ms. Modern earth leakage relays can achieve this and one such method is to use a unique sensitive polarized release as illustrated in Figure C.19.

C.5.1   Construction

The device consists of a U-shaped stator on top of which sits an armature. The magnet mounted adjacent to one limb sets up a flux strong enough to hold the armature closed against the action of the spring. There is a multi-turn coil on the other limb, which is connected to the core balance CT. When an earth fault occurs, an output is generated by the core balance CT into the coil. This reduces the standing flux to the extent that the spring takes over, to flip the armature onto the tripping bar, to open the breaker. The calibration grub screw is a magnetic shunt.

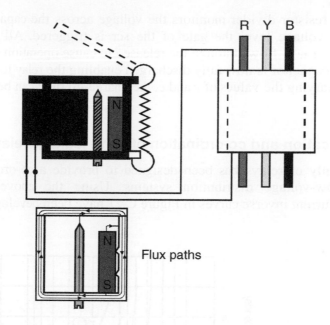

Figure C.19
IES 4 polarized release

Screwing it in bleeds off magnetism from the main loop making the release more sensitive. Screwing it out allows more magnetism around the main loop, making the armature attraction stronger, hence less sensitive. The burden of the release is only 400 micro VA (10 mV, 40 mA) which allows extremely high sensitivities to be achieved. The release can be complimented by the addition of some electronics in order to produce a series of inverse-time/current tripping curves (Figure C.20).

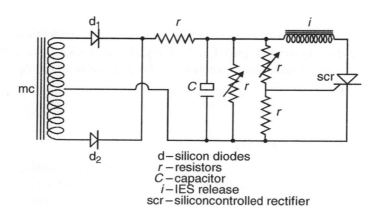

Figure C.20
Internal circuitry

C.5.2   Description of operation

When an earth fault or an earth leakage condition occurs on the system, the core balance CT mc generates an output. On the positive half cycle, the secondary current flows through the diode d_1, resistor r, and charges up capacitor c. On the negative half cycle, the current flows through the diode d_2, resistor r and charges up capacitor c even further.

The resistor divider monitors the voltage across the capacitor c and once it reaches a preset voltage level the gate of the scr is triggered. All of the energy stored in the capacitor now flows through the release i to cause operation of the relay.

The capacitor is now fully discharged enabling the relay to be reset immediately.

By varying the values of r and c, the charge-up time can be varied.

C.5.3   Application and co-ordination of earth leakage relays

A family of relays has been designed to provide a co-ordinated earth fault protection for low-voltage distribution systems. Using the above-mentioned technology, the time/current inverse curves in Figure C.21 have been developed.

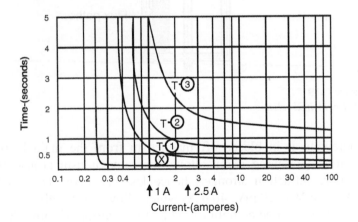

Figure C.21
Time/current response curves

This allows co-ordinated sensitive earth fault protection to be applied to a typical distribution system (Figure C.22). They afford 'backup' protection to the end relay, which provides instantaneous protection to the apparatus where operators are most likely to be working.

C.5.4   Optimum philosophy

It is important to note that the choice of relay settings cannot be considered in isolation. They are influenced by the manner of neutral earthing, current pickup levels, and time grading intervals, which will in turn be dictated by the system configuration.

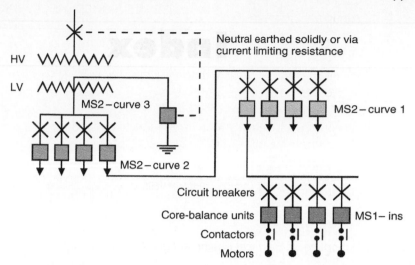

Figure C.22
Typical LV distribution system

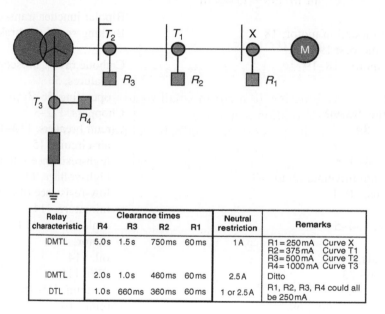

Relay characteristic	Clearance times				Neutral restriction	Remarks
	R4	R3	R2	R1		
IDMTL	5.0 s	1.5 s	750 ms	60 ms	1 A	R1 = 250 mA Curve X R2 = 375 mA Curve T1 R3 = 500 mA Curve T2 R4 = 1000 mA Curve T3
IDMTL	2.0 s	1.0 s	460 ms	60 ms	2.5 A	Ditto
DTL	1.0 s	660 ms	360 ms	60 ms	1 or 2.5 A	R1, R2, R3, R4 could all be 250 mA

Figure C.23
Optimum philosophy

All are inter-dependent, and in Figure C.23, it will be seen that an optimum philosophy for the system would be a 'definite time lag' philosophy (DTL) as opposed to an 'inverse definite time lag' philosophy (IDMTL) as faster clearance times can be achieved.

Index